Mathematical Modeling

Mathematical Modeling

Mark M. Meerschaert

Associate Professor
Department of Mathematics
Albion College
  and
Visiting Associate Professor
Department of Statistics and Probability
Michigan State University

ACADEMIC PRESS, INC.
Harcourt Brace Jovanovich, Publishers
Boston   San Diego   New York
London   Sydney   Tokyo   Toronto

Copyright © 1993 by Academic Press, Inc.

ACADEMIC PRESS, INC.
1250 Sixth Avenue, San Diego, CA 92101–4311

United Kingdom Edition published by
ACADEMIC PRESS LIMITED
24–28 Oval Road, London NW1 7DX

Library of Congress Cataloging-in-Publication Data

Meerschaert, Mark M.
    Mathematical modeling / Mark M. Meerschaert.
        p.   cm.
    Includes bibliographical references and index.
    ISBN 0-12-487650-1
        1. Mathematical models.   I. Title
QA401.M48   1993
511'.8–dc20   92-38059
                                                            CIP

Printed in the United States of America

93 94 95 96 EB 9 8 7 6 5 4 3 2 1

This book is dedicated to my wife, Carmen,
and to my two sons, Bob and Steven.
Your love and support are all the success
I will ever need.

Acknowledgments

I would like to thank Chuck Glaser, Joe Clifford, and Cindy Kogut at Academic Press for all of their help in preparing this manuscript. Thanks also to Annie Todd for bringing the book to the attention of Academic Press, to my secretary Diane Hines for her excellent typing, and to all of the reviewers for their insightful comments. Thanks to Peter Cherry and everyone at Vector Research, Inc., for teaching me about real-world problems. Thanks to Ron Fryxell and all of my colleagues at Albion College for giving me so much support in my efforts to teach others about real-world problems. Most of all, thanks to my students for cheerfully participating in the classroom experiments that eventually evolved into this textbook. To the extent that the book is a success, most of the credit belongs to you.

The text was prepared using TeX version 2.0.0.

3-D graphs were produced using AXUM by Trimetrix, Inc.

All other computer-generated graphs were produced using MICRO-SOFT WORKS version 2.00 by Microsoft, Inc.

Contents

Preface

This text, which is intended to serve as a general introduction to the area of mathematical modeling, is aimed at advanced undergraduate or beginning graduate students in mathematics and closely related fields. Formal prerequisites consist of the usual freshman–sophomore sequence in mathematics, including one-variable calculus, multivariable calculus, linear algebra, and differential equations. Prior exposure to computing and probability and statistics is useful, but is not required.

Unlike some textbooks that focus on one kind of mathematical model, this book covers the broad spectrum of modeling problems, from optimization to dynamical systems to stochastic processes. Unlike some other textbooks that assume knowledge of only a semester of calculus, this book challenges students to use *all* of the mathematics they know (because that is what it takes to solve real problems).

The overwhelming majority of mathematical models fall into one of three categories: optimization models; dynamic models; and probability models. The type of model used in a real application might be dictated by the problem at hand, but more often it is a matter of choice. In many instances, more than one type of model will be used. For example, a large Monte Carlo simulation model may be used in conjunction with a smaller, more tractable deterministic model based on expected values.

This book is organized into three parts, corresponding to the three main categories of mathematical models. We begin with optimization models. A five-step method for mathematical modeling is introduced in Section 1 of Chapter 1, in the context of one-variable optimization problems. The remainder of the first chapter is an introduction to sensitivity analysis and robustness. These fundamentals of mathematical modeling are used in a consistent way throughout the rest of the book. Exercises at the end of each chapter require students to master them as well. Chapter 2, on multivariable optimization, introduces decision variables, feasible and optimal solutions, and constraints. A review of the method of Lagrange multipliers is provided for the benefit of those students who were not exposed to this important technique in multivariable calculus. In the section on sensitivity analysis for problems with constraints, we learn that Lagrange multipliers represent shadow prices (some authors call them dual variables). This sets the stage for our discussion of linear programming later in Chapter 3. Chapter 3 covers some important computational techniques, including Newton's method in one and several variables, and linear programming.

In the next part of the book, on dynamic models, students are introduced to the concepts of state and equilibrium. Later discussions of state space, state variables, and equilibrium for stochastic processes are intimately connected to what is done here. Nonlinear dynamical systems in both discrete and continuous time are covered. There is very little emphasis on exact analytical solutions in this part of the book, since most of these models admit no analytic solution.

Finally, in the last part of the book, we introduce probability models. No prior knowledge of probability is assumed. Instead we build upon the material in the first two parts of the book, to introduce probability in a natural and intuitive way as it relates to real-world problems.

Each chapter in this book is followed by a set of challenging exercises. These exercises require significant effort on the part of the student, as well as a certain amount of creativity. I did not invent the problems in this book. They are real problems. They were not designed to illustrate the use of any particular mathematical technique. Quite the opposite. We will occasionally go over some new mathematical techniques in this book *because the problem demands it.* I was determined that there would be no place in this book where a student could look up and ask, "What is all of this for?" Although typically oversimplified or grossly unrealistic, story problems embody the fundamental challenge in applying mathematics to solve real problems. For most students, story problems present plenty of challenge. This book teaches students how to solve story problems. There is a general method that can be applied successfully by any reasonably capable student to solve any story problem. It appears in Section 1, Chapter 1. This same general method is applied to problems of all kinds throughout the text.

Following the exercises in each chapter is a list of suggestions for further reading. This list includes references to a number of UMAP modules in applied mathematics which are relevant to the material in the chapter. UMAP modules can provide interesting supplements to the material in the text, or extra credit projects. All of the UMAP modules are available at a nominal cost from COMAP, Inc., 60 Lowell Street, Arlington VA 02174.

One of the major themes of this book is the use of appropriate technology for solving mathematical problems. Computer algebra systems, graphics, and numerical methods all have their place in mathematics. Many students have not had an adequate introduction to these tools. In this course we introduce modern technology in context. Students are motivated to learn because the new technology provides a more convenient way to solve real-world problems. Computer algebra systems and 2-D graphics are useful throughout the course. Some 3-D graphics are used in Chapters 2 and 3 in the sections on multivariable optimization. Students who have already been introduced to 3-D graphics should be encouraged to use what they know. Numerical methods covered in the text include, among others, Newton's method, linear programming, the Euler method, and linear regression.

The text contains numerous computer-generated graphs, along with instruction on the appropriate use of graphing utilities in mathematics. Computer algebra systems are used extensively in those chapters where significant algebraic calculation is required. The text includes actual computer output from the computer algebra systems MAPLE and MATHEMATICA in Chapters 2, 4, 5, and 8. The chapters on computational techniques (Chapters 3, 6, and 9) discuss the appropriate use of numerical algorithms to solve problems that admit no analytic solution. Section 3.3 on linear programming includes actual computer output from the popular linear programming package LINDO. Section 8.3 on linear regression includes output from the commonly used statistical package MINITAB.

Students need to be provided with access to appropriate technology in order to take full advantage of this textbook. We have tried to make it easy for instructors to use this textbook at their own institution, whatever their situation. Some will have the means to provide students with access to sophisticated computing facilities, while others will have to make do with less. The bare necessities include: (1) a software utility to draw 2-D graphs; and (2) a machine on which students can execute a few simple numerical algorithms. All of this can be done, for example, with a computer spreadsheet program or a programmable graphics calculator. The ideal situation would be to provide all students access to a good computer algebra system (including graphics), a linear programming package, and a statistical computing package. The end of this preface provides information on some of the most popular computer software packages for the benefit of those students or instructors who do not have access to the appropriate software at their own institution.

The numerical algorithms in the text are presented in the form of pseudocode. Some instructors will prefer to have students implement the algorithms on their own. On the other hand, if students are not going to be required to program, we want to make it easy for instructors to provide them with appropriate software. All of the algorithms in the text have been implemented on a variety of computer platforms that can be made available to users of this textbook at no additional cost.

xiv   Preface

If you are interested in obtaining a copy, please contact the author. Also, if you are willing to share your own implementation with other instructors and students, please send us a copy. With your permission we will make copies available to others at no charge.

Mathematical modeling is the link between mathematics and the rest of the world. You ask a question. You think a bit, and then you refine the question, phrasing it in precise mathematical terms. Once the question becomes a mathematics question, you use mathematics to find an answer. Then finally (and this is the part that too many people forget), you have to reverse the process, translating the mathematical solution back into a comprehensible, no-nonsense answer to the original question. Some people are fluent in English, and some people are fluent in calculus—we have plenty of each. We need more people who are fluent in both languages and are willing and able to translate. These are the people who will be influential in solving the problems of the future.

Software

The following is a partial list of appropriate software packages that can be used in conjunction with this textbook.

Computer Algebra Systems

DERIVE, Soft Warehouse, Inc., 3615 Harding Ave., Suite 505, Honolulu HI 96816–3735. MAPLE, Waterloo Maple Software, 160 Columbia Street West, Waterloo, Ontario, Canada N2L 3L3, or Brooks/Cole Publishing Company, BE962, 511 Forest Lodge Rd., Pacific Grove CA 93950. MATHEMATICA, Wolfram Research, Inc., PO Box 6059, Champaign IL 61826-9902.

Statistical Packages

BDMP, BDMP Statistical Software, Inc., 1440 Sepulveda Blvd., Suite 316, Los Angeles CA 90025, (800) 238–BDMP. MINITAB, Minitab, Inc., 3081 Enterprise Dr., State College PA 16801-3008, (800) 448–3555. SAS, SAS Institute, Inc., (919) 677–8000. SPSS, SPSS Inc., 444 N. Michigan Ave., Chicago IL 60611, (312) 329–3500. JMP, Systat, Inc., 1800 Sherman Ave., Evanston IL 60201.

Linear Programming Packages

LINDO, The Scientific Press, 651 Gateway Blvd., Suite 1100, South San Francisco CA 94080–7014, (800) 451–5409. MICROSOLVE, Holden–Day, 4432 Telegraph Ave., Oakland CA 94609. TURBO–SIMPLEX, Maximal Software, Klapparas 11, IS–110 Reykjavik, Iceland.

I

Optimization Models

Problems in optimization are the most common applications of mathematics. Whatever the activity in which we are engaged, we want to maximize the good that we do and minimize the unfortunate consequences or costs. Business managers attempt to control variables in order to maximize profit or to achieve a desired goal for production and delivery at a minimum cost. Managers of renewable resources such as fisheries and forests try to control harvest rates in order to maximize long-term yield. Government agencies set standards to minimize the environmental costs of producing consumer goods. Computer system managers try to maximize throughput and minimize delays. Farmers space their plantings to maximize yield. Physicians regulate medications to minimize harmful side-effects. What all of these applications and many more have in common is a particular mathematical structure. One or more variables can be controlled to produce the best outcome in some other variable, subject in most cases to a variety of practical constraints on the control variables. Optimization models are designed to determine the values of the control variables which lead to the optimal outcome, given the constraints of the problem.

Chapter One

One-Variable Optimization

We begin our discussion of optimization models at a place where most students will already have some practical experience. One-variable optimization problems, sometimes called maximum–minimum problems, are typically discussed in first-semester calculus. A wide variety of practical applications can be handled using just these techniques. The purpose of this chapter, aside from a review of these basic techniques, is to introduce the fundamentals of mathematical modeling in a familiar setting.

1.1   The Five-Step Method

In this section we outline a general procedure that can be used to solve problems using mathematical modeling. We will illustrate this procedure, called the *five-step method*, by using it to solve a one-variable maximum–minimum problem typical of those encountered by most students in the first semester of calculus.

Example 1.1   A pig weighing 200 pounds gains 5 pounds per day and costs 45 cents a day to keep. The market price for pigs is 65 cents per pound, but is falling 1 cent per day. When should the pig be sold?

The mathematical modeling approach to problem solving consists of five steps:

1. Ask the question.

2. Select the modeling approach.

3. Formulate the model.

4. Solve the model.

5. Answer the question.

The first step is to ask a question. The question must be phrased in mathematical terms, and it often requires a good deal of work to do this. In the process we are required to make a number of assumptions or suppositions about the way things really are. We should not be afraid to make a guess at this stage. We can always come back and make a better guess later on. Before we can ask a question in mathematical terms we need to define our terms. Go through the problem and make a list of variables. Include appropriate units. Next make a list of assumptions about these variables. Include any relations between variables (equations and inequalities) that are known or assumed. Having done all of this, we are ready to ask a question. Write down in explicit mathematical language the objective of this problem. Notice that the preliminary steps of listing variables, units, equations and inequalities, and other assumptions are really a part of the question. They frame the question.

In Example 1.1 the weight w of the pig (in lbs), the number of days t until we sell the pig, the cost C of keeping the pig t days (in dollars), the market price p for pigs ($/lb), the revenue R obtained when we sell the pig ($), and our resulting net profit P ($) are all variables. There are other numerical quantities involved in the problem, such as the initial weight of the pig (200 lbs). However, these are not variables. It is important at this stage to separate variables from those quantities that will remain constant.

Next we need to list our assumptions about the variables identified in the first stage of step 1. In the process we will take into account the effect of the constants in the problem. The weight of the pig starts at 200 lbs and goes up by 5 lbs/day so we have

$$(w \text{ lbs}) = (200 \text{ lbs}) + \left(\frac{5 \text{ lbs}}{\text{day}}\right)(t \text{ days}).$$

Notice that we have included units as a check that our equation makes sense. The

Variables:
t = time (days)
w = weight of pig (lbs)
p = price for pigs ($/lb)
C = cost of keeping pig t days ($)
R = revenue obtained by selling pig ($)
P = profit from sale of pig ($)

Assumptions:
$w = 200 + 5t$
$p = 0.65 - 0.01t$
$C = 0.45t$
$R = p \cdot w$
$P = R - C$
$t \geq 0$

Objective: Maximize P

Figure 1.1   Results of step 1 of the pig problem.

other assumptions inherent in our problem are as follows:

$$\left(\frac{p \text{ dollars}}{\text{lb}} \right) = \left(\frac{0.65 \text{ dollars}}{\text{lb}} \right) - \left(\frac{0.01 \text{ dollars}}{\text{lb} \cdot \text{day}} \right) (t \text{ days})$$

$$(C \text{ dollars}) = \left(\frac{0.45 \text{ dollars}}{\text{day}} \right) (t \text{ days})$$

$$(R \text{ dollars}) = \left(\frac{p \text{ dollars}}{\text{lb}} \right) (w \text{ lbs})$$

$$(P \text{ dollars}) = (R \text{ dollars}) - (C \text{ dollars})$$

We are also assuming that $t \geq 0$. Our objective in this problem is to maximize our net profit, P dollars. Figure 1.1 summarizes the results of step 1, in a form convenient for later reference.

The three stages of step 1 (variables, assumptions, and objective) need not be completed in any particular order. For example, it is often useful to determine the objective early in step 1. In Example 1.1, it may not be readily apparent that R and C are variables until we have defined our objective, P, and we recall that $P = R - C$. One way to check that step 1 is complete is to see whether our objective P relates all the way back to the variable t. The best general advice about step 1 is to *do something*. Start by writing down whatever is immediately apparent (e.g., some of the variables can be found simply by reading over the problem and looking for nouns), and the rest of the pieces will probably fall into place.

Step 2 is to select the modeling approach. Now that we have a question stated in mathematical language, we need to select a mathematical approach to use to get an answer. Many types of problems can be stated in a standard form for which an effective general solution procedure exists. Most research in applied mathematics consists of identifying these general categories of problems and inventing efficient ways to solve them. There is a considerable body of literature in this area, and many new advances continue to be made. Few, if any, students in this course will have the experience and familiarity with the literature to make a good selection for the modeling approach. In this book, with rare exceptions, problems will specify the modeling approach to be used. Our example problem will be modeled as a one-variable optimization problem, or maximum–minimum problem.

We outline the modeling approach we have selected. For complete details we refer the reader to any introductory calculus textbook.

> We are given a real-valued function $y = f(x)$ defined on a subset S of the real line. There is a theorem that states that if f attains its maximum or minimum at an interior point $x \in S$, then $f'(x) = 0$, assuming that f is differentiable at x. This allows us to rule out any interior point $x \in S$ at which $f'(x) \neq 0$ as a candidate for max/min. This procedure works well as long as there are not too many exceptional points.

Step 3 is to formulate the model. We need to take the question exhibited in step 1 and reformulate it in the standard form selected in step 2, so that we can apply the standard general solution procedure. It is often convenient to change variable names if we will refer to a modeling approach that has been described using specific variable names, as is the case here. We write

$$P = R - C$$
$$= p \cdot w - 0.45t$$
$$= (0.65 - 0.01t)(200 + 5t) - 0.45t.$$

Let $y = P$ be the quantity we wish to maximize and $x = t$ the independent variable. Our problem now is to maximize

$$y = f(x)$$
$$= (0.65 - 0.01x)(200 + 5x) - 0.45x \tag{1}$$

over the set $S = \{x : x \geq 0\}$.

Step 4 is to solve the model, using the standard solution procedure identified in step 2. In our example we want to find the maximum of the function $y = f(x)$

defined by Eq. (1) over the set $x \geq 0$. Figure 1.2 shows a graph of the function $f(x)$. Since f is quadratic in x, the graph is a parabola. We compute that

$$f'(x) = (8 - x)/10,$$

so that $f'(x) = 0$ at the point $x = 8$. Since f is increasing on the interval $(-\infty, 8)$ and decreasing on $(8, \infty)$ the point $x = 8$ is the global maximum. At this point we have $y = f(8) = 133.20$. Since the point $(x, y) = (8, 133.20)$ is the global maximum of f over the entire real line, it is also the maximum over the set $x \geq 0$.

Step 5 is to answer the question posed originally in step 1—when to sell the pig in order to maximize profit. The answer obtained by our mathematical model is to sell the pig after eight days, thus obtaining a net profit of \$133.20. This answer is valid as long as the assumptions made in step 1 remain valid. Related questions and alternative assumptions can be addressed by changing what we did in step 1. Since we are dealing with a real problem (A farmer owns pigs. When should they be sold?), there is an element of risk involved in step 1. For that reason it is usually necessary to investigate several alternatives. This process, called *sensitivity analysis*, will be discussed in the next section.

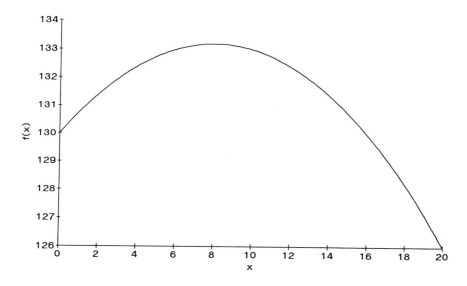

Figure 1.2 Graph of net profit $f(x)$ versus time to sell x for the pig problem.

$$f(x) = (0.65 - 0.01x)(200 + 5x) - 0.45x$$

The main purpose of this section was to introduce the five-step method for mathematical modeling. Figure 1.3 summarizes the method in a form convenient for later reference. In this book we will apply the five-step method to solve a wide variety of problems in mathematical modeling. Our discussion of step 2 will generally include a description of the modeling approach selected, along with an example or two. The reader who is already familiar with the modeling approach may choose to skip this part, or just skim to pick up the notation. Some of the other points summarized in Fig. 1.3, such as the use of "appropriate technology," will be expanded upon later in this book.

The exercises at the end of each chapter also require the application of the five-step method. Getting in the habit of using the five-step method now will make it easier to succeed on the more difficult modeling problems to come. Be sure to pay particular attention to step 5. In the real world, it is not enough to be right. You also need the ability to communicate your findings to others, some of whom may not be as mathematically knowledgeable as you.

1.2 Sensitivity Analysis

The preceding section outlines the five-step approach to mathematical modeling. The process begins by making some assumptions about the problem. We are rarely certain enough about things to be able to expect all of these assumptions to be exactly valid. Therefore, we need to consider how sensitive our conclusions are to each of the assumptions we have made. This kind of sensitivity analysis is an important aspect of mathematical modeling. The details vary according to the modeling approach used, and so our discussion of sensitivity analysis will continue throughout this book. We will focus here on sensitivity analysis for simple one-variable optimization problems.

In the preceding section we used the pig problem (Example 1.1) to illustrate the five-step approach to mathematical modeling. Figure 1.1 summarizes the assumptions we made in solving that problem. In this instance the data and assumptions were mostly spelled out for us. Even so, we need to be critical. Data are obtained by measurement, observation, and sometimes sheer guess. We need to consider the possibility that the data are not precise.

Some data are naturally known with much more certainty than others. The current weight of the pig, the current price for pigs, and the cost per day of keeping the pig are easy to measure and are known to a great degree of certainty. The rate of growth of the pig is a bit less certain, and the rate at which the price is falling is even less certain. Let r denote the rate at which the price is falling. We assumed that $r = 0.01$ dollars per day, but let us now suppose that the actual value of r is

Step 1. Ask the question.

> (a) Make a list of all the variables in the problem, including appropriate units.
> (b) Be careful not to confuse variables and constants.
> (c) State any assumptions you are making about these variables, including equations and inequalities.
> (d) Check units to make sure that your assumptions make sense.
> (e) State the objective of the problem in precise mathematical terms.

Step 2. Select the modeling approach.

> (a) Choose a general solution procedure to be followed in solving this problem.
> (b) Generally speaking, success in this step requires experience, skill, and familiarity with the relevant literature.
> (c) In this book we will usually specify the modeling approach to be used.

Step 3. Formulate the model.

> (a) Restate the question posed in step 1 in the terms of the modeling approach specified in step 2.
> (b) You may need to relabel some of the variables specified in step 1 in order to agree with the notation used in step 2.
> (c) Note any additional assumptions made in order to fit the problem described in step 1 into the mathematical structure specified in step 2.

Step 4. Solve the model.

> (a) Apply the general solution procedure specified in step 2 to the specific problem formulated in step 3.
> (b) Be careful in your mathematics. Check your work for math errors. Does your answer make sense?
> (c) Use appropriate technology. Computer algebra systems, graphics, and numerical software will increase the range of problems within your grasp, and they also help reduce math errors.

Step 5. Answer the question.

> (a) Rephrase the results of step 4 in nontechnical terms.
> (b) Avoid mathematical symbols and jargon.
> (c) Anyone who can understand the statement of the question as it was presented to you should be able to understand your answer.

Figure 1.3 The five-step method.

Table I.   Sensitivity of best time to sell x to rate r
at which price is falling for the pig problem.

r (\$/day)	x (days)
0.008	15.0
0.009	11.1
0.010	8.0
0.011	5.5
0.012	3.3

different. By repeating the solution procedure for several different values of r we can get an idea of the sensitivity of our answer to the value of r. Table I shows the results of solving our problem for a few selected values of r. Figure 1.4 contains the same sensitivity data in graphical form. We can see that the optimal time to sell is quite sensitive to the parameter r.

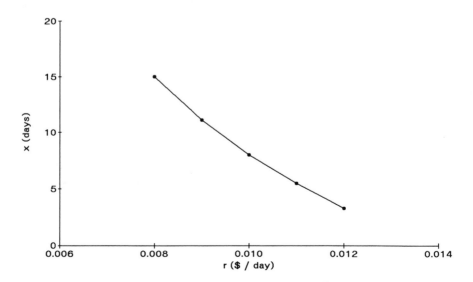

Figure 1.4 Graph of best time to sell x versus rate r at which price is falling for the pig problem.

A more systematic method for measuring this sensitivity would be to treat r as an unknown parameter, following the same steps as before. Writing

$$p = 0.65 - rt,$$

we can proceed as before to obtain

$$y = f(x)$$
$$= (0.65 - rx)(200 + 5x) - 0.45x.$$

Then we can compute

$$f'(x) = -2(25rx + 500r - 7)/5$$

so that $f'(x) = 0$ at the point

$$x = (7 - 500r)/25r. \qquad (2)$$

The optimal time to sell is given by Eq. (2) as long as $x \geq 0$, i.e., as long as $0 < r \leq 0.14$. For $r > 0.14$, the vertex of the parabola $y = f(x)$ lies outside of

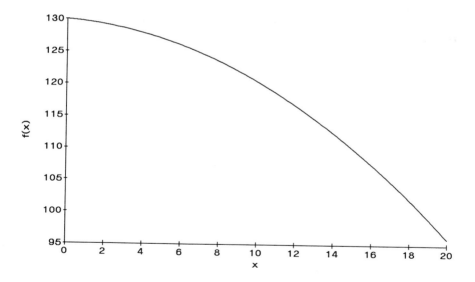

Figure 1.5 Graph of net profit $f(x)$ versus time to sell x for the pig problem in the case $r = 0.15$.

$$f(x) = (0.65 - 0.015x)(200 + 5x) - 0.45x$$

the set $x \geq 0$ over which we are maximizing. In this case the optimal time to sell is at $x = 0$ since we have $f' < 0$ on the entire interval $[0, \infty)$. See Figure 1.5 for an illustration in the case $r = 0.15$.

We are also uncertain about the growth rate g of the pig. We have assumed that $g = 5$ lbs/day. More generally, we have that

$$w = 200 + gt,$$

which leads to the equation

$$f(x) = (0.65 - 0.01x)(200 + gx) - 0.45x,$$

so that

$$f'(x) = -(2gx + 5(49 - 13g))/100.$$

Now $f'(x) = 0$ at the point

$$x = 5(13g - 49)/2g. \tag{3}$$

The optimal time to sell is given by Eq. (3) so long as it represents a nonnegative value of x. Figure 1.6 shows the relationship between the growth rate g and the optimal time to sell.

It is most natural and most useful to interpret sensitivity data in terms of relative change or percent change, rather than in absolute terms. For example, a

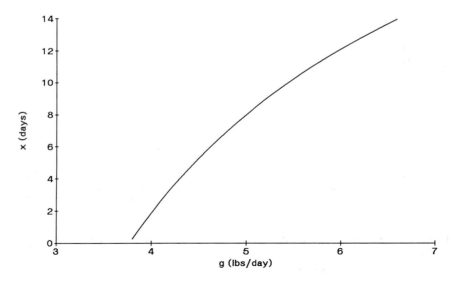

Figure 1.6 Graph of best time to sell x versus growth rate g for the pig problem.

20% decrease in r leads to a 48% increase in x, while a 10% decrease in g leads to a 34% decrease in x. If x changes by an amount Δx, the relative change in x is given by $\Delta x / x$, and the percent change in x is $100 \, \Delta x / x$. If r changes by Δr, resulting in the change Δx in x, then the ratio between the relative changes is $\Delta x / x$ divided by $\Delta r / r$. Letting $\Delta r \to 0$ and using the definition of the derivative, we obtain

$$\frac{\Delta x / x}{\Delta r / r} \to \frac{dx}{dr} \cdot \frac{r}{x}.$$

We call this limiting quantity the *sensitivity* of x to r, and we will denote it by $S(x, r)$. In the pig problem we have

$$\frac{dx}{dr} = \frac{-7}{25r^2}$$
$$= -2,800$$

at the point $r = 0.01$, thus

$$S(x, r) = \frac{dx}{dr} \cdot \frac{r}{x}$$
$$= (-2,800) \left(\frac{.01}{8} \right)$$
$$= -7/2.$$

If r goes up by 2%, then x goes down by 7%. Since

$$\frac{dx}{dg} = \frac{245}{2g^2}$$
$$= 4.9,$$

we have

$$S(x, g) = \frac{dx}{dg} \cdot \frac{g}{x}$$
$$= (4.9) \left(\frac{5}{8} \right)$$
$$= 3.0625,$$

so that a 1% increase in the growth rate of the pig would cause us to wait about 3% longer to sell the pig.

The successful application of sensitivity analysis procedures requires good judgment. It is usually not possible to compute sensitivity coefficients for each parameter in the model, nor is this particularly desirable. We need to select those parameters about which there is the most uncertainty and perform sensitivity analysis on them. The interpretation of sensitivity coefficients also depends on the degree of uncertainty, the fundamental question being the extent to which our uncertainty about the data affects our confidence in the answer. In the pig problem, we are probably considerably more certain of the growth rate g than of the rate r at which prices fall. A 25% error in g would be quite surprising if we have observed the past history of growth in this pig or in similar animals. A 25% error in our estimate of r would not be at all surprising.

1.3   Sensitivity and Robustness

A mathematical model is robust if the conclusions it leads to remain true even though the model is not completely accurate. In real problems we will never have perfect information, and even if it were possible to construct a perfectly accurate model, we might be better off with a simpler and more tractable approximation. For this reason a consideration of robustness is a necessary ingredient in any mathematical modeling project.

In the preceding section we introduced the process of sensitivity analysis, which is a way to gauge the robustness of a model with respect to assumptions about the data. There are other assumptions made in step 1 of the mathematical modeling process which should also be examined. While it is often necessary to make assumptions for purposes of mathematical convenience and simplicity, it is the responsibility of the modeler to see to it that these assumptions are not so specialized as to invalidate the results of the modeling process.

Figure 1.1 contains a summary of the assumptions made in solving the pig problem. Aside from data values, the main assumptions are that both the weight of the pig and the selling price per pound are linear functions of time. These are obviously simplifying assumptions and cannot be expected to hold exactly. After all, according to these assumptions, a year from now the pig would weigh

$$w = 200 + 5(365)$$
$$= 2,025 \, \text{lbs}$$

and sell for

$$p = 0.65 - 0.01(365)$$
$$= -3.00 \, \text{dollars/lb.}$$

A more realistic model would take into account both the nonlinearity of these functions and the increasing uncertainty as time goes on.

How can a model give the right answer if the assumptions are wrong? While mathematical modeling strives for perfection, perfection can never be achieved. It would be more descriptive to say that mathematical modeling strives *toward* perfection. A well-constructed mathematical model will be robust, which is to say that while the answers it gives may not be perfectly correct, they will be close enough to be useful in a real-world context.

Let us examine the linearity assumptions made in the pig problem. Our basic equation is

$$P = pw - 0.45t,$$

where p is the selling price of the pig in dollars per pound, and w is the weight of the pig in pounds. If the original data and assumptions of the model are not too far off, then the best time to sell the pig is obtained by setting $P' = 0$. Calculate to find

$$p'w + pw' = 0.45$$

dollars per day. The term $p'w + pw'$ represents the rate of increase in the value of the pig. Our model tells us to keep the pig as long as the value of the pig is increasing faster than the cost of feeding it. Furthermore, the change in the pig's value has two components, $p'w$ and pw'. The first term, $p'w$, represents the loss in value due to a drop in price. The second term, pw', represents the gain in value due to the pig gaining weight. Consider the practical problems involved in the application of this more general model. The data required include a complete specification of both the future growth of the pig and the future changes in price as differentiable functions of time. There is no way to know these functions exactly. There is even some question as to whether they make sense. Can the pig be sold at 3 A.M. Sunday morning? Can price be an irrational number? Let us construct a realistic scenario. The farmer has a pig weighing approximately 200 lbs. The pig has been gaining about five lbs/day during the last week. Five days ago the pig could have been sold for 70 cents/lb but by now the price has dropped to 65 cents/lb. What should we do? The obvious approach is to project on the basis of this data ($w = 200$, $w' = 5$, $p = 0.65$, $p' = -0.01$) and decide when to sell. This is exactly what we did. We understand that p' and w' will not remain constant over the next few weeks, and that therefore p and w will not be linear functions of time. However, as long as p' and w' do not change too much over this period, the error involved in assuming they remain constant will not be too great.

We are now prepared to give a somewhat broader interpretation to the results of our sensitivity analysis from the preceding section. Recall that the sensitivity of the best time to sell (x) to changes in the growth rate w' was calculated to be 3. Suppose that in fact the growth rate over the next few weeks is somewhere between 4.5 and 5.5 lbs/day. This is within 10% of the assumed value. Then the best time to sell the pig will be within 30% of 8 days, or between 5 and 11 days. The amount of lost profit by selling at 8 days is less than 1 dollar.

With regard to price, suppose that we feel the value $p' = -.01$, or a 1 cent/day drop in price over the next few weeks, is a worst-case scenario. Prices are likely to drop more slowly in the future and may even level off ($p' = 0$). All we can really say now is that we should wait at least 8 days to sell. For small values of p' (near zero), our model suggests waiting a very long time to sell. However, our model is not valid over long time intervals. The best course of action in this case is probably to keep the pig for a week, reestimate the parameter values p, p', w', and w, and start over.

1.4   Exercises

1. An automobile manufacturer makes a profit of $1,500 on the sale of a certain model. It is estimated that for every $100 of rebate, sales increase by 15%.
    (a) What amount of rebate will maximize profit? Use the five-step method, and model as a one-variable optimization problem.
    (b) Compute the sensitivity of your answer to the 15% assumption. Consider both the amount of rebate and the resulting profit.
    (c) Suppose that rebates actually generate only a 10% increase in sales per $100. What is the effect? What if the response is somewhere between 10 and 15% per $100 of rebate?
    (d) Under what circumstances would a rebate offer cause a reduction in profit?
2. In the pig problem, perform a sensitivity analysis based on the cost per day of keeping the pig. Consider both the effect on the best time to sell and on the resulting profit. If a new feed costing 60 cents/day would let the pig grow at a rate of 7 lbs/day, would it be worth switching feed? What is the minimum improvement in growth rate that would make this new feed worthwhile?
3. Reconsider the pig problem of Example 1.1, but now assume that the price for pigs is starting to level off. Let

$$p = 0.65 - 0.01t + 0.00004t^2 \qquad (4)$$

represent the price for pigs (cents/lb) after t days.
    (a) Graph Eq. (4) along with our original price equation. Explain why our original price equation could be considered as an approximation to Eq. (4) for values of t near zero.
    (b) Find the best time to sell the pig. Use the five-step method, and model as a one-variable optimization problem.
    (c) The parameter 0.00004 represents the rate at which price is leveling off. Conduct a sensitivity analysis on this parameter. Consider both the optimal time to sell and the resulting profit.

(d) Compare the results of part (b) to the optimal solution contained in the text. Comment on the robustness of our assumptions about price.

4. An oil spill has fouled 200 miles of Pacific shoreline. The oil company responsible has been given 14 days to clean up the shoreline, after which a fine will be levied in the amount of $10,000/day. The local clean-up crew can scrub 5 miles of beach per week at a cost of $500/day. Additional crews can be brought in at a cost of $18,000 plus $800/day for each crew.

(a) How many additional crews should be brought in to minimize the total cost to the company? Use the five-step method. How much will the clean-up cost?

(b) Examine the sensitivity to the rate at which a crew can clean up the shoreline. Consider both the optimal number of crews and the total cost to the company.

(c) Examine the sensitivity to the amount of the fine. Consider the number of days the company will take to clean up the spill and the total cost to the company.

(d) The company has filed an appeal on the grounds that the amount of the fine is excessive. Assuming that the only purpose of the fine is to motivate the company to clean up the oil spill in a timely manner, is the fine excessive?

5. It is estimated that the growth rate of the fin whale population (per year) is $rx(1 - x/K)$, where $r = 0.08$ is the intrinsic growth rate, $K = 400,000$ is the maximum sustainable population, and x is the current population, now around 70,000. It is further estimated that the number of whales harvested per year is about $.00001\ Ex$, where E is the level of fishing effort in boat-days. Given a fixed level of effort, population will eventually stabilize at the level where growth rate equals harvest rate.

(a) What level of effort will maximize the sustained harvest rate? Model as a one-variable optimization problem using the five-step method.

(b) Examine the sensitivity to the intrinsic growth rate. Consider both the optimum level of effort and the resulting population level.

(c) Examine the sensitivity to the maximum sustainable population. Consider both the optimum level of effort and the resulting population level.

6. In problem 5, suppose that the cost of whaling is $500 per boat-day, and the price of a fin whale carcass is $6,000.

(a) Find the level of effort that will maximize profit over the long term. Model as a one-variable optimization problem using the five-step method.

(b) Examine the sensitivity to the cost of whaling. Consider both the eventual profit in $/year and the level of effort.

(c) Examine the sensitivity to the price of a fin whale carcass. Consider both profit and level of effort.

(d) Over the past 30 years there have been several unsuccessful attempts to ban whaling worldwide. Examine the economic incentives for whalers to continue harvesting. In particular, determine the conditions (values of the two parameters: cost per boat-day and price per fin whale carcass) under which harvesting the fin whale produces a sustained profit over the long term.

7. Reconsider the pig problem of Example 1.1, but now suppose that our objective is to maximize our profit rate ($/day). Assume that we have already owned the pig for 90 days and have invested $100 in this pig to date.

(a) Find the best time to sell the pig. Use the five-step method, and model as a one-variable optimization problem.

(b) Examine the sensitivity to the growth rate of the pig. Consider both the best time to sell and the resulting profit rate.

(c) Examine the sensitivity to the rate at which the price for pigs is dropping. Consider both the best time to sell and the resulting profit rate.

8. Reconsider the pig problem of Example 1.1, but now take into account the fact that the growth rate of the pig decreases as the pig gets older. Assume that the pig will be fully grown in another five months.

(a) Find the best time to sell the pig in order to maximize profit. Use the five-step method, and model as a one-variable optimization problem.

(b) Examine the sensitivity to the time it will take until the pig is fully grown. Consider both the best time to sell and the resulting profit.

9. A local daily newspaper with a circulation of 80,000 subscribers is thinking of raising its subscription price. Currently the price is $1.50 per week, and it is estimated that the paper would lose 5,000 subscribers if the rate were to be raised by 10 cents/week.

(a) Find the subscription price that maximizes profit. Use the five-step method, and model as a one-variable optimization problem.

(b) Examine the sensitivity of your answer in part (a) to the assumption of 5,000 lost subscribers. Calculate the optimal subscription rate assuming that this parameter is 3,000, 4,000, 5,000, 6,000, or 7,000.

(c) Let $n = 5,000$ denote the number of subscribers lost when the subscription price increases by 10 cents. Calculate the optimal subscription price p as a function of n, and use this formula to determine the sensitivity $S(p, n)$.

(d) Should the paper change its subscription price? Justify your conclusions in plain English.

Further Reading

Cameron, D., Giordano, F. and Weir, M. *Modeling Using the Derivative: Numerical and Analytic Solutions*, UMAP module 625.

Cooper, L. and Sternberg, D. (1970). *Introduction to Methods of Optimization*, W. B. Saunders, Philadelphia.

Gill, P., Murray, W. and Wright, M. (1981). *Practical Optimization*, Academic Press, New York.

Meyer, W. (1984). *Concepts of Mathematical Modeling*, McGraw–Hill, New York.

Rudin W. (1976). *Principles of Mathematical Analysis*, 3rd Ed., McGraw–Hill, New York.

Whitley, W. *Five Applications of Max–Min Theory from Calculus*, UMAP module 341.

Chapter Two

Multivariable Optimization

Many optimization problems require the simultaneous consideration of a number of independent variables. In this chapter we consider the simplest category of multivariable optimization problems. The techniques should be familiar to most students from multivariable calculus. In this chapter we also introduce the use of computer algebra systems to handle some of the more complicated algebraic computations.

2.1   Unconstrained Optimization

The simplest type of multivariable optimization problems involves finding the maximum or minimum of a differentiable function of several variables over a nice set. Further complications arise, as we will see later, when the set over which we optimize takes a more complex form.

Example 2.1   A manufacturer of color TV sets is planning the introduction of two new products, a 19-inch stereo color set with a manufacturer's suggested retail price (MSRP) of $339 and a 21-inch stereo color set with an MSRP of $399. The cost to the company is $195 per 19-inch set and $225 per 21-inch set, plus an additional $400,000 in fixed costs. In the competitive market in which these sets will be sold, the number of sales per year will affect the average selling price. It is estimated that for each type of set, the average selling price drops by one cent for

Variables: s = number of 19-inch sets sold (per year)
t = number of 21-inch sets sold (per year)
p = selling price for 19-inch sets ($)
q = selling price for 21-inch sets ($)
C = cost of manufacturing sets ($/year)
R = revenue from the sale of sets ($/year)
P = profit from the sale of sets ($/year)

Assumptions: $p = 339 - .01s - .003t$
$q = 399 - .004s - .01t$
$R = ps + qt$
$C = 400,000 + 195s + 225t$
$P = R - C$
$s \geq 0$
$t \geq 0$

Objective: Maximize P

Figure 2.1 Results of step 1 of the color TV problem.

each additional unit sold. Furthermore, sales of the 19-inch set will affect sales of the 21-inch set, and vice–versa. It is estimated that the average selling price for the 19-inch sets will be reduced by an additional 0.3 cents for each 21-inch set sold, and the price for 21-inch sets will decrease by 0.4 cents for each 19-inch set sold. How many units of each type of set should be manufactured?

The five-step approach to mathematical modeling, introduced in the preceding chapter, will be used to solve this problem. Step 1 is to ask a question. We begin by making a list of variables. Next we write down the relations between variables and any other assumptions, such as nonnegativity. Finally, we formulate a question in mathematical terms, using the established notation. The results of step 1 are summarized in Figure 2.1.

Step 2 is to select the modeling approach. We will solve this problem as a multivariable unconstrained optimization problem. This type of problem is typically treated in introductory courses in multivariable calculus. We will outline the model and the general solution procedure here. We refer the reader to any introductory calculus textbook for details and mathematical proofs.

We are given a function $y = f(x_1, \ldots, x_n)$ on a subset S of the n-dimensional space $\mathbb{R}^n$. We wish to find the maximum and/or minimum values of f on the set S. There is a theorem that states that if f attains its maximum or minimum at an interior point $(x_1, \ldots, x_n)$ in S, then $\nabla f = 0$ at that point, assuming that f is differentiable at that point. In other words, at the extreme point

$$\frac{\partial f}{\partial x_1}(x_1, \ldots, x_n) = 0$$

$$\frac{\partial f}{\partial x_n}(x_1, \ldots, x_n) = 0$$

(1)

The theorem allows us to rule out as a candidate for max/min any point in the interior of S for which any of the partial derivatives of f do not equal zero. Thus, to find the max/min points we must solve simultaneously the n equations in n unknowns defined by Equation (1). Then we must also check any points on the boundary of S, as well as any points where one or more of the partial derivatives is undefined.

Step 3 is to formulate the model, using the standard form chosen in step 2. Let

$$\begin{aligned} P &= R - C \\ &= ps + qt - (400,000 + 195s + 225t) \\ &= (339 - .01s - .003t)s \\ &\quad + (399 - .004s - .01t)t \\ &\quad - (400,000 + 195s + 225t). \end{aligned}$$

Now let $y = P$ be the quantity we wish to maximize, and let $x_1 = s$, $x_2 = t$ be our decision variables. Our problem now is to maximize

$$\begin{aligned} y &= f(x_1, x_2) \\ &= (339 - .01x_1 - .003x_2)x_1 \\ &\quad + (399 - .004x_1 - .01x_2)x_2 \\ &\quad - (400,000 + 195x_1 + 225x_2) \end{aligned}$$

(2)

over the set

$$S = \{(x_1, x_2) : x_1 \geq 0, x_2 \geq 0\}.$$

(3)

Step 4 is to solve the problem, using the standard solution methods outlined in step 2. The problem is to maximize the function f given by Eq. (2) over the set

S defined in Eq. (3). Figure 2.2 shows a graph of the function f. This function is a paraboloid, and the vertex of the paraboloid is the unique solution to Eq. (1) obtained by setting $\nabla f = 0$. We compute that

$$\frac{\partial f}{\partial x_1} = 144 - .02x_1 - .007x_2 = 0$$

$$\frac{\partial f}{\partial x_2} = 174 - .007x_1 - .02x_2 = 0 \tag{4}$$

at the point

$$x_1 = \frac{554,000}{117} \approx 4,735$$

$$x_2 = \frac{824,000}{117} \approx 7,043. \tag{5}$$

The point (x_1, x_2) defined by Eq. (5) is the vertex of the paraboloid, so it represents the global maximum of f over the entire real plane. It is therefore also

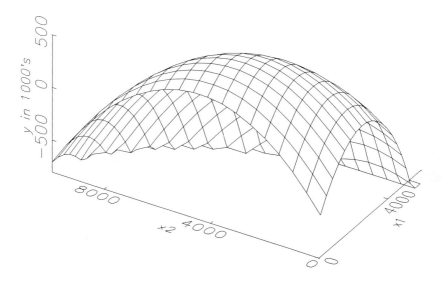

Figure 2.2 Graph of profit $y = f(x_1, x_2)$ versus production levels x_1 of 19-inch sets and x_2 of 21-inch sets for the color TV problem.

$$f(x_1, x_2) = (339 - .01x_1 - .003x_2)x_1$$
$$+ (399 - .004x_1 - .01x_2)x_2$$
$$- (400,000 + 195x_1 + 225x_2)$$

the maximum of f over the set S defined in Eq. (3). The maximum value of f is obtained by substituting Eq. (5) back into Eq. (2), which yields

$$y = \frac{21,592,000}{39} \approx 553,641. \tag{6}$$

The calculations of step 4 in this problem are a bit cumbersome. In cases like this one it is appropriate to use a *computer algebra system* to perform the necessary calculations. Computer algebra systems can differentiate, integrate, solve equations, and simplify algebraic expressions. Most packages can also perform matrix algebra, draw graphs, and solve some differential equations. Several good computer algebra systems (MAPLE, MATHEMATICA, DERIVE, etc.) are available for both mainframe and personal computers, and many systems offer a student version at a substantially reduced price. Computer algebra systems are an example of the kind of "appropriate technology" we referred to in Fig. 1.3 above, in our summary of the five-step method. Figure 2.3 shows the results of using the computer algebra system MATHEMATICA to solve the current model. There are several advantages to using a computer algebra system for a problem like this. It is more efficient and more accurate. You will be more productive if you learn to use this technology, and it will give you the freedom to concentrate more on the larger issues of problem solving instead of getting bogged down in the calculations. We will illustrate the use of computer algebra systems again in our sensitivity analysis calculations below, where the algebra is even more exacting.

The final step, step 5, is to answer the question in plain English. Simply stated, the company can maximize profits by manufacturing 4,735 of the 19-inch sets and 7,043 of the 21-inch sets, resulting in a net profit of $553,641 for the year. The average selling price for 19-inch sets is $270.52 and $309.63 for 21-inch sets. The cost of manufacture is $2,908,000, resulting in a profit margin of 19 percent. These figures indicate a profitable venture, so we would recommend that the company proceed with the introduction of these new products.

The conclusions of the preceding paragraph are based on the assumptions illustrated in Fig. 2.1. Before reporting our findings to the company, we should perform sensitivity analysis to insure that our conclusions are robust with respect to our assumptions about both the market for TV sets and the manufacturing process. Our main concern is the value of the decision variables x_1 and x_2, since the company must act on this information.

We illustrate the procedure for sensitivity analysis by examining the sensitivity to price elasticity for 19-inch sets, which we will denote by the variable a. In our model we assumed that $a = 0.01$ dollars per set. Substituting into our previous

```
In[1]:= y=(339-x1/100-3 x2/1000) x1 +
        (399-4 x1/1000-x2/100) x2 -
        (400000+195 x1+225 x2)
```

$$
Out[1]= -400000 - 195\ x1 + x1\ (339 - \frac{x1}{100} - \frac{3\ x2}{1000}) - 225\ x2 +
$$

$$
(399 - \frac{x1}{250} - \frac{x2}{100})\ x2
$$

```
In[2]:= dydx1=D[y,x1]
```

$$
Out[2]= 144 - \frac{x1}{50} - \frac{7\ x2}{1000}
$$

```
In[3]:= dydx2=D[y,x2]
```

$$
Out[3]= 174 - \frac{7\ x1}{1000} - \frac{x2}{50}
$$

```
In[4]:= s=Solve[{dydx1==0,dydx2==0},{x1,x2}]
```

$$
Out[4]= \{\{x1 \to \frac{554000}{117},\ x2 \to \frac{824000}{117}\}\}
$$

```
In[5]:= N[s]
```

```
Out[5]= {{x1 -> 4735.04, x2 -> 7042.74}}
```

```
In[6]:= y/.s
```

$$
Out[6]= \{\frac{21592000}{39}\}
$$

```
In[7]:= N[%]
```

```
Out[7]= {553641.}
```

Figure 2.3 Optimal solution to the color TV problem using the computer algebra system MATHEMATICA.

formula, we obtain

$$
\begin{aligned}
y &= f(x_1, x_2) \\
&= (339 - ax_1 - .003x_2)x_1 \\
&\quad + (399 - .004x_1 - .01x_2)x_2 \\
&\quad - (400,000 + 195x_1 + 225x_2).
\end{aligned}
\tag{7}
$$

When we compute partial derivatives and set them equal to zero, we obtain

$$
\begin{aligned}
\frac{\partial f}{\partial x_1} &= 144 - 2ax_1 - .007x_2 = 0 \\
\frac{\partial f}{\partial x_2} &= 174 - .007x_1 - .02x_2 = 0.
\end{aligned}
\tag{8}
$$

Solving for x_1 and x_2 as before yields

$$
\begin{aligned}
x_1 &= \frac{1,662,000}{40,000a - 49} \\
x_2 &= 8,700 - \frac{581,700}{40,000a - 49}
\end{aligned}
\tag{9}
$$

See Figs. 2.4 and 2.5 for the graphs of x_1 and x_2 versus a. From these graphs it appears that a higher price elasticity a for 19-inch sets will reduce the optimal production level x_1 for 19-inch sets and increase the optimal production level x_2 for 21-inch sets. It also appears that x_1 is more sensitive to a than x_2, which seems to make sense. To get a numerical measure of these sensitivities we compute

$$
\begin{aligned}
\frac{dx_1}{da} &= \frac{-66,480,000,000}{(40,000a - 49)^2} \\
&= \frac{-22,160,000,000}{41,067}
\end{aligned}
$$

at $a = 0.01$, so that

$$
\begin{aligned}
S(x_1, a) &= \left(\frac{-22,160,000,000}{41,067}\right)\left(\frac{.01}{554,000/117}\right) \\
&= -\frac{400}{351} \approx -1.1.
\end{aligned}
$$

A similar calculation yields

$$
S(x_2, a) = \frac{9,695}{36,153} \approx 0.27.
$$

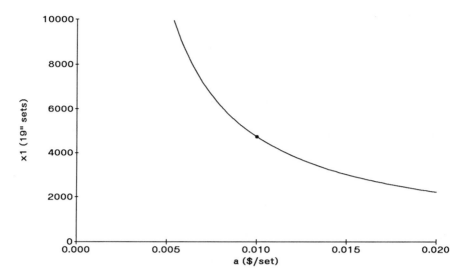

Figure 2.4 Graph of optimum production level x_1 of 19-inch sets versus price elasticity a for the color TV problem.

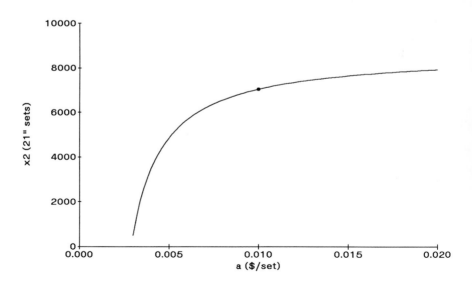

Figure 2.5 Graph of optimum production level x_2 of 21-inch sets versus price elasticity a for the color TV problem.

If the price elasticity for 19-inch sets were to increase by 10%, then we should make 11% fewer 19-inch sets and 2.7% more 21-inch sets.

Next we consider the sensitivity of y to a. What effect will a change in the price elasticity for 19-inch sets have on our profits? To obtain a formula for y in terms of a, we substitute Eq. (9) back into Eq. (7) to get

$$
\begin{aligned}
y = &\left[339 - a\left(\frac{1,662,000}{40,000a - 49} \right) - .003\left(8,700 - \frac{581,700}{40,000a - 49} \right) \right] \\
&\times \left(\frac{1,662,000}{40,000a - 49} \right) \\
&+ \left[399 - .004\left(\frac{1,662,000}{40,000a - 49} \right) - 01\left(8,700 - \frac{581,700}{40,000a - 49} \right) \right] \\
&\times \left(8,700 - \frac{581,700}{40,000a - 49} \right) \\
&- \left[400,000 + 195\left(\frac{16,620,000}{40,000a - 49} \right) + 225\left(8,700 - \frac{581,700}{40,000a - 49} \right) \right].
\end{aligned}
$$

$$(10)$$

See Fig. 2.6 for a graph of y versus a. It appears that an increase in price elasticity for 19-inch sets will result in lower profits.

To compute $S(y, a)$ we will need to obtain a formula for dy/da. One way to do this would be to employ one-variable methods directly on Eq. (10), perhaps with

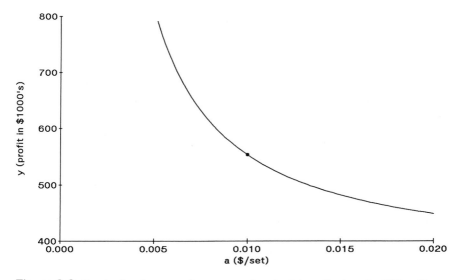

Figure 2.6 Graph of optimum profit y versus price elasticity a for the color TV problem.

the aid of a computer algebra system. Another method, which is computationally more efficient, is to use the multivariable chain rule, which implies that

$$\frac{dy}{da} = \frac{\partial y}{\partial x_1}\frac{dx_1}{da} + \frac{\partial y}{\partial x_2}\frac{dx_2}{da} + \frac{\partial y}{\partial a}. \tag{11}$$

Since both $\partial y/\partial x_1$ and $\partial y/\partial x_2$ are zero at the optimum, we have

$$\frac{dy}{da} = \frac{\partial y}{\partial a} = -x_1^2$$

directly from Eq. (2), so

$$S(y, a) = -\left(\frac{554,000}{117}\right)^2 \frac{.01}{(21,592,000/39)}$$

$$= -\frac{383,645}{947,349} \approx -0.40.$$

Thus, a 10% increase in price elasticity for 19-inch sets will result in a 4% drop in profit.

The fact that the term

$$\frac{\partial y}{\partial x_1}\frac{dx_1}{da} + \frac{\partial y}{\partial x_2}\frac{dx_2}{da} = 0$$

in Eq. (11) has its own real-world significance. This part of the derivative dy/da represents the effect on profits of the changing optimal production levels x_1 and x_2. The fact that it sums to zero means that small changes in production levels have (at least in the linear approximation) no effect on profits. Geometrically, since we are at the maximum point where the curve $y = f(x_1, x_2)$ is flat, small changes in x_1 and x_2 have little effect on y. Almost all of the drop in optimal profits caused by a 10% increase in price elasticity for 19-inch sets is due to the change in selling price. Therefore, the production levels given by our model will be very nearly optimal. For example, suppose that we have assumed $a = 0.01$, but that this price elasticity is in fact 10% higher. We will set our production levels using Eq. (5), which means that we will produce 11% too many 19-inch sets and around 3% too few 21-inch sets, compared to the optimal solution given by Eq. (9) with $a = 0.011$. Also, our profits will be 4% lower than expected. But what have we actually lost by applying the results of our model? Using Eq. (5) with $a = 0.011$, we will net a \$531,219 profit. The optimal profit would be \$533,514 (set $a = 0.011$ in Eq. (9) and substitute back into Eq. (7)). Hence, we have lost only 0.43 percent of the potential maximum profit by applying the results of our model, even though our actual production levels were quite a way off from the optimum values. Our model appears to be extremely robust in this regard.

Furthermore, a similar conclusion should hold for many similar problems, since it is basically due to the fact that $\nabla f = 0$ at a critical point.

All of the previous sensitivity analysis calculations could also be performed using a computer algebra system. In fact, this is the preferred method, assuming that one is available. Figure 2.7 illustrates how the computer algebra system MAPLE can be used to compute the sensitivity $S(x_1, a)$. The calculations of the other sensitivities are similar.

Sensitivity analysis for the other elasticities could be carried out in the same manner. While the particulars will differ, the form of the function f suggests that each affects y in essentially the same manner. In particular, we have a high degree of confidence that our model will lead to a good (nearly optimal) decision about production levels even in the presence of small errors in the estimation of price elasticities.

We will say just a few words on the more general subject of robustness. Our model is based on a linear price structure. Certainly this is only an approximation. However, in practical applications we are likely to proceed as follows. We begin with an educated guess about the size of the market for our new products and with a reasonable average sale price. Then we estimate elasticities either on the basis of past experience with similar situations or on the basis of limited marketing studies. We should be able to get reasonable estimates for these elasticities over a certain range of sales levels. This range presumably includes the optimal levels. So in effect we are simply making a linear approximation of a nonlinear function over a fairly small region. This sort of approximation is well known to exhibit robustness. After all, this is the whole idea behind calculus.

2.2 Lagrange Multipliers

In this section we begin to consider optimization problems with a more sophisticated structure. As we noted at the beginning of the previous section, complications arise in the solution of multivariable optimization models when the set over which we optimize becomes more complex. In real problems we are led to consider these more complicated models by the existence of constraints on the independent variables.

Example 2.2 We reconsider the color TV problem (Example 2.1) introduced in the previous section. There we assumed that the company has the potential to produce any number of TV sets per year. Now we will introduce constraints based

```
> y:=(339-a*x1-3*x2/1000)*x1
>    +(399-4*x1/1000-x2/100)*x2-(400000+195*x1+225*x2);
    y := (339 - a x1 - 3/1000 x2) x1 + (399 - 1/250 x1 -
1/100 x2) x2


            - 400000 - 195 x1 - 225 x2


> dydx1:=diff(y,x1);
                    dydx1 := - 2 a x1 + 144 - 7/1000 x2


> dydx2:=diff(y,x2);
                    dydx2 := - 7/1000 x1 - 1/50 x2 + 174


> s:=solve({dydx1=0,dydx2=0},{x1,x2});
                        - 21 + 7250 a              1662000
        s := {x2 = 48000 --------------, x1 = --------------}
                        - 49 + 40000 a          - 49 + 40000 a


> assign(s);
> dx1da:=diff(x1,a);

                              66480000000
                    dx1da := - ----------------
                                            2
                              (- 49 + 40000 a)


> assign(a=1/100);
> x1;

                              554000
                              ------
                               117

> sx1a:=dx1da*(a/x1);
                                    400
                        sx1a := - ---
                                    351


> evalf(sx1a);
                        -1.139601140
> stop;
```

Figure 2.7 Calculation of the sensitivity $S(x_1, a)$ for the color TV problem using the computer algebra system MAPLE.

on the available production capacity. Consideration of these two new products came about because the company plans to discontinue manufacture of its black-and-white sets, thus providing excess capacity at its assembly plant. This excess capacity could be used to increase production of other existing product lines, but the company feels that the new products will be more profitable. It is estimated that the available production capacity will be sufficient to produce 10,000 sets per year ($\approx$ 200 per week). The company has an ample supply of 19-inch and 21-inch color picture tubes, chassis, and other standard components; however, the circuit boards necessary for constructing a stereo TV set are currently in short supply. Also, the 19-inch TV requires a different board than the 21-inch model because of the internal configuration, which cannot be changed without a major redesign, which the company is not prepared to undertake at this time. The supplier is able to supply 8,000 boards per year for the 21-inch model and 5,000 boards per year for the 19-inch model. Taking this information into account, how should the company set production levels?

Once again we will employ the five-step method. The results of step 1 are shown in Figure 2.8. The only change is the addition of several constraints on the decision variables s and t. Step 2 is to select the modeling approach. This problem will be modeled as a multivariable constrained optimization problem and solved using the method of Lagrange multipliers.

We are given a function $y = f(x_1, \ldots, x_n)$ and a set of constraints. For the moment we will assume that these constraints can be expressed in the form of k functional equations

$$g_1(x_1, \ldots, x_n) = c_1$$
$$g_2(x_1, \ldots, x_n) = c_2$$
$$\vdots$$
$$g_k(x_1, \ldots, x_n) = c_k.$$

Later on we will explain how to handle inequality constraints. Our job is to optimize

$$y = f(x_1, \ldots, x_n)$$

over the set

$$S = \{(x_1, \ldots, x_n) : \text{all } g_i(x_1, \ldots, x_n) = c_i \text{ for } i = 1, \ldots, k\}.$$

There is a theorem that states that at an extreme point $x \in S$, we must have

$$\nabla f = \lambda_1 \nabla g_1 + \cdots + \lambda_k \nabla g_k.$$

We call $\lambda_1, \ldots, \lambda_k$ the *Lagrange multipliers*. This theorem assumes that $\nabla g_1, \ldots, \nabla g_k$ are linearly independent vectors (Edwards, (1973), p. 113).

Variables:

s = number of 19-inch sets sold (per year)

t = number of 21-inch sets sold (per year)

p = selling price for 19-inch sets (\$)

q = selling price for 21-inch sets (\$)

C = cost of manufacturing sets (\$/year)

R = revenue from the sale of sets (\$/year)

P = profit from the sale of sets (\$/year)

Assumptions:

$p = 339 - .01s - .003t$

$q = 399 - .004s - .01t$

$R = ps + qt$

$C = 400,000 + 195s + 225t$

$P = R - C$

$s \leq 5000$

$t \leq 8000$

$s + t \leq 10,000$

$s \geq 0$

$t \geq 0$

Objective: Maximize P

Figure 2.8 Results of step 1 for the color TV problem with constraints.

Then in order to locate the max/min points of f on the set S, we must solve the n Lagrange multiplier equations:

$$\frac{\partial f}{\partial x_1} = \lambda_1 \frac{\partial g_1}{\partial x_1} + \cdots + \lambda_k \frac{\partial g_k}{\partial x_1}$$

$$\vdots$$

$$\frac{\partial f}{\partial x_n} = \lambda_1 \frac{\partial g_1}{\partial x_n} + \cdots + \lambda_k \frac{\partial g_k}{\partial x_n}$$

together with the k constraint equations:

$$g_1(x_1, \ldots, x_n) = c_1$$

$$\vdots$$

$$g_k(x_1, \ldots, x_n) = c_k$$

for the variables $x_1, \ldots, x_n$ and $\lambda_1, \ldots, \lambda_k$. We must also check any exceptional points at which the gradient vectors $\nabla g_1, \ldots, \nabla g_k$ are not linearly independent.

The method of Lagrange multipliers is based on a geometrical interpretation of the gradient vector. Suppose for the moment that there is only one constraint equation,

$$g(x_1, \ldots, x_n) = c,$$

so that the Lagrange multiplier equation becomes

$$\nabla f = \lambda \nabla g.$$

The set $g = c$ is a curved surface of dimension $n - 1$ in $\mathbb{R}^n$, and for any point $x \in S$ the gradient vector $\nabla g(x)$ is perpendicular to S at that point. The gradient vector ∇f always points in the direction in which f increases the fastest. At a local max or min, the direction in which f increases fastest must also be perpendicular to S, so at that point we must have ∇f and ∇g pointing along the same line, i.e., $\nabla f = \lambda \nabla g$.

In the case of several constraints, the geometrical argument is similar. Now the set S represents the intersection of the k level surfaces $g_1 = c_1, \ldots, g_k = c_k$. Each one of these is an $(n - 1)$-dimensional subset of $\mathbb{R}^n$, so their intersection is an $(n - k)$-dimensional subset. At an extreme point, ∇f must be perpendicular to the set S. Therefore it must lie in the space spanned by the k vectors $\nabla g_1, \ldots, \nabla g_k$. The technical condition of linear independence ensures that the k vectors $\nabla g_1, \ldots, \nabla g_k$ actually do span a k-dimensional space. (In the case of a single constraint, linear independence simply means that $\nabla g \neq 0$.)

Example 2.3   Maximize $x + 2y + 3z$ over the set $x^2 + y^2 + z^2 = 3$.

This is a constrained multivariable optimization problem. Let

$$f(x, y, z) = x + 2y + 3z$$

denote the objective function, and let

$$g(x, y, z) = x^2 + y^2 + z^2$$

denote the constraint function. Compute

$$\nabla f = (1, 2, 3)$$
$$\nabla g = (2x, 2y, 2z).$$

At the maximum, $\nabla f = \lambda \nabla g$, in other words,

$$1 = 2x\lambda$$
$$2 = 2y\lambda$$
$$3 = 2z\lambda.$$

This gives three equations in four unknowns. Solving in terms of λ, we obtain

$$x = 1/2\lambda$$
$$y = 1/\lambda$$
$$z = 3/2\lambda.$$

Using the fact that
$$x^2 + y^2 + z^2 = 3,$$

we obtain a quadratic equation in λ, with two real roots. The root $\lambda = \sqrt{42}/6$ leads to

$$x = 1/2\lambda = \sqrt{42}/14$$
$$y = 1/\lambda = \sqrt{42}/7$$
$$z = 3/2\lambda = 3\sqrt{42}/14,$$

so that the point

$$a = (\sqrt{42}/14, \ \sqrt{42}/7, \ 3\sqrt{42}/14)$$

is one candidate for the maximum. The other root, $\lambda = -\sqrt{42}/6$, leads to another candidate, $b = -a$. Since $\nabla g \neq 0$ everywhere on the constraint set $g = 3$, a and b are the only two candidates for the maximum. Since f is a continuous function on the closed and bounded set $g = 3$, f must attain its maximum and minimum on this set. Then, since

$$f(a) = \sqrt{42}, \qquad\qquad f(b) = -\sqrt{42},$$

the point a is the maximum and b is the minimum. Consider the geometry of this example. The constraint set S defined by the equation

$$x^2 + y^2 + z^2 = 3$$

is a sphere of radius $\sqrt{3}$ centered at the origin in $\mathbb{R}^3$. Level sets of the objective function

$$f(x, \ y, \ z) = x + 2y + 3z$$

are planes in R^3. The points a and b are the only two points on the sphere S at which one of these planes is tangent to the sphere. At the maximum point a, the gradient vectors ∇f and ∇g point in the same direction. At the minimum point b, ∇f and ∇g point in opposite directions.

Example 2.4   Maximize $x + 2y + 3z$ over the set $x^2 + y^2 + z^2 = 3$ and $x = 1$.

The objective function is

$$f(x, \ y, \ z) = x + 2y + 3z,$$

so

$$\nabla f = (1, 2, 3).$$

The constraint functions are

$$g_1(x, y, z) = x^2 + y^2 + z^2$$
$$g_2(x, y, z) = x.$$

Compute

$$\nabla g_1 = (2x, 2y, 2z)$$
$$\nabla g_2 = (1, 0, 0).$$

Then the Lagrange multiplier formula $\nabla f = \lambda_1 \nabla g_1 + \lambda_2 \nabla g_2$ yields

$$1 = 2x\lambda_1 + \lambda_2$$
$$2 = 2y\lambda_1$$
$$3 = 2z\lambda_1.$$

Solving for x, y, and z in terms of λ_1 and λ_2 gives

$$x = (1 - \lambda_2)/2\lambda_1$$
$$y = 2/2\lambda_1$$
$$z = 3/2\lambda_1.$$

Substituting into the constraint equation $x = 1$ gives $\lambda_2 = 1 - 2\lambda_1$. Substituting all of this into the remaining equation

$$x^2 + y^2 + z^2 = 3$$

yields a quadratic equation in λ_1, which gives $\lambda_1 = \pm\sqrt{26}/4$. Substituting back into the equations for x, y, and z yields the two following solutions:

$$c = (1, 2\sqrt{26}/13, 3\sqrt{26}/13)$$
$$d = (1, -2\sqrt{26}/13, -3\sqrt{26}/13).$$

Since the two gradient vectors ∇g_1 and ∇g_2 are linearly independent everywhere on the constraint set, the points c and d are the only candidates for a maximum. Since f must attain its maximum on this closed and bounded set, we need only evaluate f at each candidate point to find the maximum. The maximum is

$$f(c) = 1 + \sqrt{26},$$

and the point d is the location of the minimum. The constraint set S in this example is a circle in $\mathbb{R}^3$ formed by the intersection of the sphere

$$x^2 + y^2 + z^2 = 3$$

and the plane $x = 1$. As before, the level sets of the function f are planes in $\mathbb{R}^3$. At the points c and d these planes are tangent to the circle S.

Inequality constraints can be handled by a combination of the Lagrange multiplier technique and the techniques for unconstrained problems. Suppose that the problem in Example 2.4 is altered by replacing the $x = 1$ constraint with the inequality constraint $x \leq 1$. We can consider the set

$$S = \{(x,\ y,\ z) : x^2 + y^2 + z^2 = 3,\ x \leq 1\}$$

as the union of two components. The maximum over the first component

$$S_1 = \{(x,\ y,\ z) : x^2 + y^2 + z^2 = 3,\ x = 1\}$$

was found to occur at the point

$$c = (1,\ \sqrt{8/13},\ 1.5\sqrt{8/13})$$

in our previous analysis, and we can calculate that

$$f(x,\ y,\ z) = 1 + 6.5\sqrt{8/13} = 6.01$$

at this point. To consider the remaining part

$$S_2 = \{(x,\ y,\ z) : x^2 + y^2 + z^2 = 3,\ x < 1\},$$

we return to our analysis from Example 2.3, noting that there is no local maximum of f anywhere on this set. Therefore, the maximum of f on S_1 must be the maximum of the function f on the set S. If we had considered the maximum of f over the set

$$S = \{(x,y,z) : x^2 + y^2 + z^2 = 3,\ x \geq 1\},$$

then the maximum would be at the point

$$a = (1/2,\ 2/2,\ 3/2) \cdot \sqrt{6/7}$$

found in our analysis of Example 2.3.

Returning now to the problem introduced at the beginning of this section, we are ready to continue the modeling process with step 3. We will formulate the revised color TV problem as a constrained multivariable optimization problem. We wish to maximize $y = P$ (profit) as a function of our two decision variables, $x_1 = s$ and $x_2 = t$. We have the same objective function

$$\begin{aligned}
y &= f(x_1,\ x_2)\\
&= (339 - .01x_1 - .003x_2)x_1 + (399 - .004x_1 - .01x_2)x_2\\
&\quad - (400,000 + 195x_1 + 225x_2).
\end{aligned}$$

We wish to maximize f over the set S consisting of all x_1 and x_2 satisfying the constraints

$$x_1 \leq 5,000$$
$$x_2 \leq 8,000$$
$$x_1 + x_2 \leq 10,000$$
$$x_1 \geq 0$$
$$x_2 \geq 0.$$

The set S is called the *feasible region* because it represents the set of all feasible production levels. Figure 2.9 shows a graph of the feasible region for this problem.

We will apply Lagrange multiplier methods to find the maximum of $y = f(x_1, x_2)$ over the set S. Compute

$$\nabla f = (144 - .02x_1 - .007x_2, \ 174 - .007x_1 - .02x_2).$$

Since $\nabla f \neq 0$ in the interior of S, the maximum must occur on the boundary. Consider first the segment of the boundary on the constraint line

$$g(x_1, x_2) = x_1 + x_2 = 10,000.$$

Here $\nabla g = (1, 1)$, so the Lagrange multiplier equations are

$$144 - .02x_1 - .007x_2 = \lambda$$
$$174 - .007x_1 - .02x_2 = \lambda. \tag{12}$$

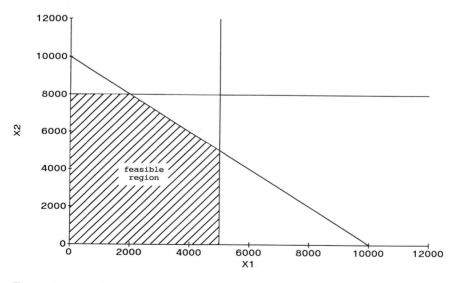

Figure 2.9 Graph showing the set of all feasible production levels x_1 of 19-inch sets and x_2 of 21-inch sets for the color TV problem with constraints.

Solving these two equations together with the constraint equation

$$x_1 + x_2 = 10,000$$

yields

$$x_1 = \frac{50,000}{13} \approx 3,846$$

$$x_2 = \frac{80,000}{13} \approx 6,154$$

$$\lambda = 24.$$

Since $y = f(x_1, x_2)$ becomes negative far away from the origin, this point represents the global maximum of f on $g = 10,000$. Furthermore, it is not hard to see that this point actually represents the maximum of f on $g \leq 10,000$, which contains the entire feasible region S. Hence we have found the maximum over S. Some of the calculations in step 4 are rather involved. In such cases it is appropriate to use a computer algebra system to simplify the process of computing derivatives and solving equations. Figure 2.10 shows the results of using the computer algebra system MATHEMATICA to perform the calculations of step 4 for this problem.

In plain English, the company can maximize profits by producing 3,846 of the 19-inch sets and 6,154 of the 21-inch sets for a total of 10,000 sets per year. This level of production uses all of the available excess production capacity. The resource constraints on the availability of stereo TV circuit boards are not binding. This venture will produce an estimated profit of $532,308 annually.

2.3 Sensitivity Analysis and Shadow Prices

In this section we discuss some of the specialized techniques for sensitivity analysis in Lagrange multiplier models. It turns out that the multipliers themselves have a real-world significance.

Before we report the results of our model analysis in Example 2.2, it is important to perform sensitivity analysis. At the end of Section 2.1 we investigated the sensitivity to price elasticity for a model without constraints. The procedure for our new model is not much different. We examine the sensitivity to a particular parameter value by generalizing the model slightly, replacing the assumed value by a variable. Suppose we want to look again at the price elasticity, a, for 19-inch sets. We rewrite the objective function as in Eq. (7) so that

$$\nabla f = \left(\frac{\partial f}{\partial x_1}, \frac{\partial f}{\partial x_2} \right),$$

```
In[1]:= y=(339-x1/100-3 x2/1000)x1+
           (399-4 x1/1000-x2/100)x2-
           (400000+195 x1+225 x2)
```

$$
\text{Out[1]= } -400000 - 195\ x1 + x1\ \left(339 - \frac{x1}{100} - \frac{3\ x2}{1000}\right) - 225\ x2 +
$$

$$
\left(399 - \frac{x1}{250} - \frac{x2}{100}\right)\ x2
$$

```
In[2]:= dydx1=D[y,x1]
```

$$
\text{Out[2]= } 144 - \frac{x1}{50} - \frac{7\ x2}{1000}
$$

```
In[3]:= dydx2=D[y,x2]
```

$$
\text{Out[3]= } 174 - \frac{7\ x1}{1000} - \frac{x2}{50}
$$

```
In[4]:= s=Solve[{dydx1==lambda,dydx2==lambda,x1+x2==10000},{x1,x2,lambda}]
```

$$
\text{Out[4]= } \left\{\left\{x1 \to \frac{50000}{13},\ x2 \to \frac{80000}{13},\ lambda \to 24\right\}\right\}
$$

```
In[5]:= N[%]
```

```
Out[5]= {{x1 -> 3846.15, x2 -> 6153.85, lambda -> 24.}}
```

```
In[6]:= y/.%
```

```
Out[6]= {532308.}
```

Figure 2.10 Optimal solution to the color TV problem with constraints using the computer algebra system MATHEMATICA.

where $\partial f/\partial x_1$ and $\partial f/\partial x_2$ are given by Eq. (8). Now the Lagrange multiplier equations are

$$144 - 2ax_1 - .007x_2 = \lambda$$
$$174 - .007x_1 - .02x_2 = \lambda. \tag{13}$$

Solving together with the constraint equation

$$g(x_1,\ x_2) = x_1 + x_2 = 10,000$$

yields

$$x_1 = \frac{50,000}{1,000a + 3}$$

$$x_2 = 10,000 - \frac{50,000}{1,000a + 3} \tag{14}$$

$$\lambda = \frac{650}{1,000a + 3} - 26.$$

Then we have

$$\frac{dx_1}{da} = \frac{-50,000,000}{(1,000a + 3)^2}$$

$$\frac{dx_2}{da} = \frac{-dx_1}{da}, \tag{15}$$

so that at the point $x_1 = 3,846$, $x_2 = 6,154$, $a = 0.01$, we obtain

$$S(x_1,\ a) = \frac{dx_1}{da} \cdot \frac{a}{x_1} = -0.77$$

$$S(x_2,\ a) = \frac{dx_2}{da} \cdot \frac{a}{x_2} = 0.48.$$

See Figures 2.11 and 2.12 for the graphs of x_1 and x_2 versus a in this case. If the price elasticity of 19-inch sets increases, we will shift production from 19-inch to 21-inch sets. If it decreases, then we will produce more 19-inch sets and fewer 21-inch sets. In any case, as long as the point $(x_1,\ x_2)$ given by Eq. (14) lies between the other constraint lines ($.007 \le a \le .022$), we will always produce a total of 10,000 sets.

Now let us consider the sensitivity of our optimal profit y to the price elasticity a for 19-inch sets. To obtain a formula for y in terms of a, we substitute Eq. (13)

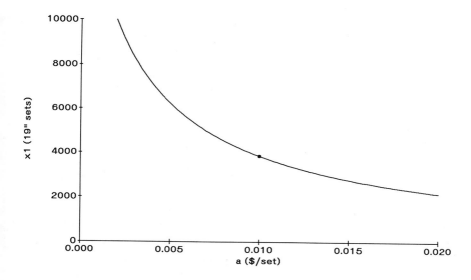

Figure 2.11 Graph of optimal production level x_1 of 19-inch sets versus price elasticity a for the color TV problem with constraints.

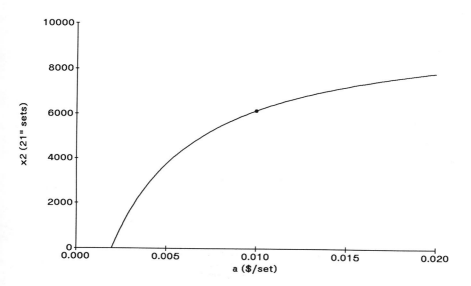

Figure 2.12 Graph of optimal production level x_2 of 21-inch sets versus price elasticity a for the color TV problem with constraints.

back into Eq. (6) to get

$$y = \left[339 - a\left(\frac{50,000}{1,000a+3} \right) - .003\left(10,000 - \frac{50,000}{1,000a+3} \right) \right]\left(\frac{50,000}{1,000a+3} \right)$$
$$+ \left[399 - .004\left(\frac{50,000}{1,000a+3} \right) - .01\left(10,000 - \frac{50,000}{1,000a+3} \right) \right]$$
$$\times \left(1,000 - \frac{50,000}{1,000a+3} \right)$$
$$- \left[400,000 + 195\left(\frac{50,000}{1,000a+3} \right) + 225\left(10,000 - \frac{50,000}{1,000a+3} \right) \right].$$

See Figure 2.13 for the graph of y versus a. In order to get a numerical measure of the sensitivity of y to a, we could apply one-variable techniques to the above expression, perhaps with the aid of a computer algebra system. Another method, which is much more efficient computationally, is to use the multivariable chain rule in Eq. (11). For any a, the gradient vector ∇f is perpendicular to the constraint line $g = 10,000$. Since

$$x(a) = (x_1(a),\ x_2(a))$$

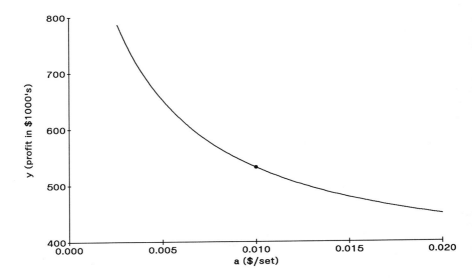

Figure 2.13 Graph of optimum profit y versus price elasticity a for the color TV problem with constraints.

is a point on the curve $g = 10,000$, the velocity vector

$$\frac{dx}{da} = \left(\frac{dx_1}{da}, \frac{dx_2}{da}\right)$$

is tangent to the curve. But then ∇f is perpendicular to dx/da, i.e., the dot product.

$$\nabla f \cdot \frac{dx}{da} = \left(\frac{\partial y}{\partial x_1}, \frac{\partial y}{\partial x_2}\right) \cdot \left(\frac{dx_1}{da}, \frac{dx_2}{da}\right)$$

$$= \frac{\partial y}{\partial x_1}\frac{dx_1}{da} + \frac{\partial y}{\partial x_2}\frac{dx_2}{da} = 0$$

in general. Therefore we again obtain

$$\frac{dy}{da} = \frac{\partial y}{\partial a} = -x^2$$

as in Section 2.1. Now we can easily compute that

$$S(y, a) = \frac{dy}{da} \cdot \frac{a}{y}$$

$$= -(3,846)^2 \frac{.01}{532,308}$$

$$= -0.28.$$

As in the unconstrained problem, an increase in price elasticity leads to lost profits. It is also true here that, as in the unconstrained problem, almost all of the lost profit is due to the fact that the selling price for 19-inch sets has decreased. If $a = 0.011$ and we use $x_1 = 3,846$, $x_2 = 6,154$ instead of the new optimum given by Eq. (13), we will not lose much potential profit. The gradient vector ∇f points in the direction of fastest increase of the objective function f, which represents our profits. We are not at the optimum, but the path between the optimum and the point (3,846, 6,154) is perpendicular to ∇f. Therefore, we may expect that the value of f at this point does not vary much from the optimum value. Hence our model leads to a nearly optimal decision even in the presence of small variations in a.

We remark that for this problem it would also make sense to use a computer algebra system to perform the necessary calculations. Figure 2.14 illustrates the use of the computer algebra system MAPLE to compute the sensitivity $S(x_2, a)$. The other sensitivities can be computed in a similar manner.

We will now consider the sensitivity of the optimal production levels x_1 and x_2 and the resulting profit y to the available manufacturing capacity of $c = 10,000$ sets per year. In order to do this we will return to our original problem, replacing the constraint $g = 10,000$ by the more general form $g = c$. The feasible region is

```
> y:=(339-a*x1-3*x2/1000)*x1
>    +(399-4*x1/1000-x2/100)*x2-(400000+195*x1+225*x2);
     y := (339 - a x1 - 3/1000 x2) x1 + (399 - 1/250 x1 - 1/100 x2) x2


           - 400000 - 195 x1 - 225 x2


> dydx1:=diff(y,x1);
                    dydx1 := - 2 a x1 + 144 - 7/1000 x2


> dydx2:=diff(y,x2);
                    dydx2 := - 7/1000 x1 - 1/50 x2 + 174


> s:=solve({dydx1=lambda,dydx2=lambda,x1+x2=10000},{x1,x2,lambda});

                    500 a - 11              - 1 + 500 a              50000
     s := {lambda = - 52 ----------, x2 = 20000 -----------, x1 = ----------}
                    3 + 1000 a              3 + 1000 a            3 + 1000 a


> assign(s);
> dx2da:=diff(x2,a);

                    10000000                  - 1 + 500 a
             dx2da := ---------- - 20000000 -------------
                    3 + 1000 a                      2
                                            (3 + 1000 a)


> assign(a=1/100);
> sx2a:=dx2da*(a/x2);

                                25
                    sx2a := ----
                                52


> evalf(sx2a);
                    .4807692308

> stop;
```

Figure 2.14 Calculation of the sensitivity $S(x_2, a)$ for the color TV problem with constraints using the computer algebra system MAPLE.

similar to that pictured in Fig. 2.9, but now the slanted constraint line is moved a bit (remaining parallel to the line $x_1 + x_2 = 10,000$). For values of c near $10,000$, the maximum still occurs at the point on the constraint line

$$g(x_1, x_2) = x_1 + x_2 = c \tag{16}$$

where $\nabla f = \lambda \nabla g$. Since both ∇f and ∇g are unchanged from our original problem, we have the same Lagrange multiplier equations from Eq. (12), which are to be solved along with the new constraint equation, Eq. (16). Solving we obtain

$$x_1 = \frac{13c - 30,000}{26}$$
$$x_2 = \frac{13c + 30,000}{26} \tag{17}$$
$$\lambda = \frac{3(106,000 - 9c)}{2,000}.$$

Now

$$\frac{dx_1}{dc} = \frac{1}{2}$$
$$\frac{dx_2}{dc} = \frac{1}{2}. \tag{18}$$

There is a simple geometric explanation for Eq. (18). Since ∇f points in the direction of the fastest increase of f, when we move the constraint line in Eq. (16), the new optimum (x_1, x_2) should be located at approximately the point where ∇f intersects the line in Eq. (16). Then

$$S(x_1, c) = \frac{1}{2} \cdot \frac{10,000}{3,846} \approx 1.3$$
$$S(x_2, c) = \frac{1}{2} \cdot \frac{10,000}{6,154} \approx 0.8.$$

To obtain the sensitivity of y to c, we compute

$$\frac{dy}{dc} = \frac{\partial y}{\partial x_1} \frac{dx_1}{dc} + \frac{\partial y}{\partial x_2} \frac{dx_2}{dc}$$
$$= (24)(\frac{1}{2}) + (24)(\frac{1}{2})$$
$$= 24,$$

which is the value of the Lagrange multiplier λ. Now

$$S(y, c) = (24)\left(\frac{10,000}{532,308}\right) \approx 0.45.$$

The geometric explanation for $dy/dc = \lambda$ is as follows. We have $\nabla f = \lambda \nabla g$, and as we increase c we move out in the direction ∇f. As we move in this direction, f increases λ times as fast as g.

The derivative $dy/dc = 24$ has an important real-world interpretation. The addition of 1 unit of production capacity $\Delta c = 1$ results in an increased profit $\Delta y = 24$ dollars. This is called a *shadow price*. It represents the value to the company of a certain resource (production capacity). If the company is interested in the possibility of increasing production capacity, which is, after all, the binding constraint, it should be willing to pay up to \$24 per unit of added capacity. Alternatively, it would be worthwhile to transfer production capacity from the manufacture of 19-inch and 21-inch stereo color TV sets to an alternative product if and only if the new product would yield a profit greater than \$24 per unit.

The calculation of sensitivities in this problem can be simplified by using a computer algebra system. Figure 2.15 illustrates the use of the computer algebra system MAPLE to compute the sensitivity $S(y, c)$. The other sensitivities can be computed in a similar manner. If you have the good fortune to have access to a computer algebra system, you should use it for your own work. Real-world problems often involve lengthy computations. Some facility in the use of a computer algebra system will make you more productive. It is also a lot more fun than calculation by hand.

Of course, the optimal level of profit y and the production levels x_1 and x_2 are totally insensitive to the values of the other constraint coefficients, since the other constraints, $x_1 \leq 5,000$ and $x_2 \leq 8,000$, are not binding. A small change in the upper bounds on x_1 or x_2 would change the feasible region, but the optimal solution would remain at (3,846, 6,154). Thus, the shadow prices for these resources are zero. The company would be unwilling to pay a premium to increase the available number of stereo TV circuit boards, since they do not need them. This situation would not change unless the number of boards available was reduced to 3,846 or less for 19-inch sets, or to 6,154 or less for 21-inch sets. In the next example, we will consider what happens in this case.

Example 2.5   Suppose that in the constrained color TV problem (Example 2.2) the number of circuit boards available for 19-inch TVs is only 3,000 per year. What is the optimum production schedule?

In this case the point on the level curve $x_1 + x_2 = 10,000$ at which $f(x_1, x_2)$ is maximized occurs outside of the feasible region. The maximum of f on the feasible region occurs at the point (3,000, 7,000). This is the intersection of the constraint curves

$$g_1(x_1, x_2) = x_1 + x_2 = 10,000$$
$$g_2(x_1, x_2) = x_1 = 3,000.$$

```
> y:=(339-x1/100-3*x2/1000)*x1
>    +(399-4*x1/1000-x2/100)*x2-(400000+195*x1+225*x2);
  y := (339 - 1/100 x1 - 3/1000 x2) x1 + (399 - 1/250 x1 - 1/100 x2) x2

          - 400000 - 195 x1 - 225 x2


> dydx1:=diff(y,x1);
                    dydx1 := - 1/50 x1 + 144 - 7/1000 x2


> dydx2:=diff(y,x2);
                    dydx2 := - 7/1000 x1 - 1/50 x2 + 174


> s:=solve({dydx1=lambda,dydx2=lambda,x1+x2=c},{x1,x2,lambda});
                     27                    15000                 15000
    s := {lambda = - ---- c + 159, x2 = 1/2 c + -----, x1 = 1/2 c - -----}
                    2000                          13                   13


> assign(s);
> dydc:=diff(y,c);

                          27
              dydc := - ---- c + 159
                        2000


assign(c=10000);
> assign(c=10000);
> dydc;

                          24


> syc:=dydc*(c/y);

                        78
               syc := ---
                       173


> evalf(syc);
                    .4508670520


> stop;
```

Figure 2.15 Calculation of the sensitivity $S(y, c)$ for the color TV problem with constraints using the computer algebra system MAPLE.

At this point we have

$$\nabla f = \lambda_1 \nabla g_1 + \lambda_2 \nabla g_2.$$

In fact we can easily compute

$$\nabla f = (35, \ 13)$$
$$\nabla g_1 = (1, \ 1)$$
$$\nabla g_2 = (1, \ 0)$$

at the point (3,000, 7,000), thus we have $\lambda_1 = 13$ and $\lambda_2 = 22$. Of course any vector in $\mathbb{R}^2$ can be written as a linear combination of $(1, 1)$ and $(1, 0)$. The point, however, of computing the Lagrange multipliers is that even in the case of multiple constraints they still represent the shadow prices for the binding constraints (production capacity and 19-inch boards). In other words, an additional unit of production capacity is worth \$13, and an additional circuit board is worth \$22.

For the convenience of the reader, we include here a proof of the fact that the Lagrange multipliers represent shadow prices. We are given a function $y = f(x_1, \ldots, x_n)$ which is to be optimized over the set defined by one or more constraint equations of the form

$$g_1(x_1, \ldots, x_n) = c_1$$
$$g_2(x_1, \ldots, x_n) = c_2$$

$$\vdots$$

$$g_k(x_1, \ldots, x_n) = c_k.$$

Suppose that the optimum occurs at a point x_0 at which the hypotheses of the Lagrange multiplier theorem are satisfied, so that at this point we have

$$\nabla f = \lambda_1 \nabla g_1 + \cdots + \lambda_k \nabla g_k. \tag{19}$$

Since the constraint equations can be written in any order, it will suffice to show that λ_1 is the shadow price corresponding to the first constraint. Let $x(t)$ denote the location of the optimum point over the set $g_1 = t, g_2 = c_2, \ldots, g_k = c_k$. Since $g_1(x(t)) = t$, we have $\nabla g(x(t)) \cdot x'(t) = 1$ and in particular $\nabla g(x_0) \cdot x'(c_1) = 1$. Since for $i = 2, \ldots, k$ we have $g_i(x(t)) = c_i$ constant for all t, we have $\nabla g_i(x(t)) \cdot x'(t) = 0$ and in particular $\nabla g_i(x_0) \cdot x'(c_1) = 0$. The shadow price is

$$d(f(x(t)))/dt = \nabla f(x(t)) \cdot x'(t)$$

evaluated at the point $t = c_1$. Then from the fact that Eq. (19) holds at the point x_0, we obtain $\nabla f(x_0) \cdot x'(c_1) = \lambda_1 \nabla g_1(x_0) \cdot x'(c_1) = \lambda_1$ as desired.

2.4 Exercises

1. Ecologists use the following model to represent the growth process of two competing species, x and y:

$$\frac{dx}{dt} = r_1 x (1 - x/K_1) - \alpha_1 xy$$

$$\frac{dy}{dt} = r_2 y (1 - y/K_2) - \alpha_2 xy.$$

The variables x and y represent the number in each population. The parameters r_i represent the intrinsic growth rates of each species; K_i represents the maximum sustainable population in the absence of competition; and α_i represents the effects of competition. Studies of the blue whale and fin whale populations have determined the following parameter values (t in years):

	Blue	Fin
r	0.05	0.08
K	150, 000	400, 000
α	10^{-8}	10^{-8}

(a) Determine the population levels x and y that maximize the number of new whales born each year. Use the five-step method, and model as an unconstrained optimization problem.

(b) Examine the sensitivity of the optimal population levels to the intrinsic growth rates r_1 and r_2.

(c) Examine the sensitivity of the optimal population levels to the environmental carrying capacities K_1 and K_2.

(d) Assuming that $\alpha_1 = \alpha_2 = \alpha$, is it ever optimal for one species to become extinct?

2. Reconsider the whale problem of Exercise 1, but now look at the total number of whales. We will say that the whale population levels x and y are feasible provided that both x and y are nonnegative. We will say that the population levels x and y are sustainable provided that both of the growth rates dx/dt and dy/dt are nonnegative.

(a) Determine the population levels that are feasible, sustainable, and that maximize the total whale population $x + y$. Use the five-step method, and model as a constrained optimization problem.

(b) Examine the sensitivity of the optimal population levels x and y to the intrinsic growth rates r_1 and r_2.

(c) Examine the sensitivity of the optimal population levels x and y to the environmental carrying capacities K_1 and K_2.

(d) Assuming that $\alpha_1 = \alpha_2 = \alpha$, examine the sensitivity of the optimal population levels x and y to the strength of competition α. Is it ever optimal to drive one species to extinction?

3. Reconsider the whale problem of Exercise 1, but now consider the economic ramifications of harvesting.

   (a) A blue whale carcass is worth $12,000, and a fin whale carcass is worth about half as much. Assuming that controlled harvesting can be used to maintain x and y at any desired level, what population levels will produce the maximum revenue? (Once population reaches the desired levels, the population levels will be kept constant by harvesting at a rate equal to the growth rate.) Use the five-step method. Model as an unconstrained optimization problem.

   (b) Examine the sensitivity of the optimal population levels x and y to the parameters r_1 and r_2.

   (c) Examine the sensitivity of revenue in $/year to the parameters r_1 and r_2.

   (d) Assuming $\alpha_1 = \alpha_2 = \alpha$, study the sensitivity of x and y to α. At what point does it become economically optimal to drive a species to extinction?

4. In Exercise 1, suppose that the International Whaling Commission (IWC) has decreed that no population of whales may be sustained at a level less than half of the environmental carrying capacity K.

   (a) Find the population levels that maximize the sustained profit subject to these constraints. Use Lagrange multipliers.

   (b) Examine the sensitivity of the optimal population levels x and y and the sustained profit to the constraint coefficients.

   (c) The IWC feels that enforcement of the minimum population rules is most easily carried out in terms of quotas. Determine the quotas (maximum number of blue whales and fin whales harvested per year) that will have an equivalent effect to the $K/2$ rule.

   (d) The whalers, complaining that IWC quotas cost them a considerable amount of money, have petitioned for them to be relaxed. Analyze the potential effects of increased quotas on both the yearly revenue of the whalers and the population levels of the whales.

5. Consider the color TV problem without constraints (Example 2.1). Because the company's assembly plant is located overseas, the U.S. government has imposed a tariff of $25 per unit.

   (a) Find the optimal production levels, taking the tariff into consideration. What does the tariff cost the company? How much of this cost is paid directly to the government, and how much represents lost sales?

   (b) Would it be worthwhile for the company to relocate production facilities to the U.S. in order to avoid the tariff? Assume that the overseas facility can be leased to another manufacturer for $200,000 per year and that the cost of constructing and operating a new facility in the U.S.

would amount to $550,000 annually. The construction costs have been amortized over the expected life of the new facility.

(c) The purpose of the tariff is to motivate manufacturing companies to operate plants in the U.S. What is the minimum tariff that would make it worthwhile for the company to relocate its facility?

(d) Given that the tariff is large enough to motivate the company to move its facility, how important is the actual tariff amount? Consider the sensitivity of both production levels and profit to the amount of the tariff.

6. A manufacturer of personal computers currently sells 10,000 units per month of a basic model. The cost of manufacture is $700/unit, and the wholesale price is $950. During the last quarter the manufacturer lowered the price $100 in a few test markets, and the result was a 50% increase in sales. The company has been advertising its product nationwide at a cost of $50,000 per month. The advertising agency claims that increasing the advertising budget by $10,000/month would result in a sales increase of 200 units/month. Management has agreed to consider an increase in the advertising budget to no more than $100,000/month.

(a) Determine the price and the advertising budget that will maximize profit. Use the five-step method. Model as a constrained optimization problem, and solve using the method of Lagrange multipliers.

(b) Determine the sensitivity of the decision variables (price and advertising) to price elasticity—the 50% number.

(c) Determine the sensitivity of the decision variables to the advertising agency's estimate of 200 new sales each time the advertising budget is increased by $10,000 per month.

(d) What is the value of the multiplier found in part (a)? What is the real-world significance of the multiplier? How could you use this information to convince top management to lift the ceiling on advertising expenditures?

7. A local daily newspaper has recently been acquired by a large media conglomerate. The paper currently sells for $1.50/week and has a circulation of 80,000 subscribers. Advertising sells for $250/page, and the paper currently sells 350 pages/week (50 pages/day). The new management is looking for ways to increase profits. It is estimated that an increase of 10 cents/week in the subscription price will cause a drop in circulation of 5,000 subscribers. Increasing the price of advertising by $100/page will cause the paper to lose approximately 50 pages of advertising per week. The loss of advertising will also affect circulation, since one of the reasons people buy the paper is for the advertisements. It is estimated that a loss of 50 pages of advertisements

per week will reduce circulation by 1,000 subscriptions.

(a) Find the weekly subscription price and the advertising price that will maximize profit. Use the five-step method, and model as an unconstrained optimization problem.

(b) Examine the sensitivity of your conclusions in part (a) to the assumption of 5,000 lost sales when the price of the paper increases by 10 cents.

(c) Examine the sensitivity of your conclusions in part (a) to the assumption of 50 pages/week of lost advertising sales when the price of advertising is increased by $100/page.

(d) Advertisers who currently place advertisements in the newspaper have the option of using direct mail to reach their customers. Direct mail would cost the equivalent of $500/page of newspaper advertising. How does this information alter your conclusions in part (a)?

8. Reconsider the newspaper problem of Exercise 7, but now suppose that advertisers have the option of using direct mail to reach their customers. Because of this, management has decided not to increase the price of advertising beyond $400/page.

(a) Find the weekly subscription price and the advertising price that will maximize profit. Use the five-step method, and model as a constrained optimization problem. Solve by the method of Lagrange multipliers.

(b) Determine the sensitivity of your decision variables (subscription price and advertising price) to the assumption of 5,000 lost sales when the price of the paper increases by 10 cents.

(c) Determine the sensitivity of the two decision variables to the assumption of 50 pages of advertisements lost per week when the advertising price increases by $100/page.

(d) What is the value of the Lagrange multiplier found in part (a)? Interpret this number in terms of the sensitivity of profit to the $400/page assumption.

9. Reconsider the newspaper problem of Exercise 7, but now look at the newspaper's business expenses. The current weekly business expenses for the paper are: $80,000 for the editorial department (news, features, editorials); $30,000 for the sales department (advertising); $30,000 for the circulation department; and $60,000 in fixed costs (mortgage, utilities, maintenance). The new management is considering cuts in the editorial department. It is estimated that the paper can operate with a minimum of a $40,000/week editorial budget. Reducing the editorial budget will save money, but it will also affect the quality of the paper. Experience in other markets suggests that the paper will lose 2% of its subscribers and 1% of its advertisers for every 10% cut in the editorial budget. Management is also considering an increase

in the sales budget. Recently the management of another paper in a similar market expanded its advertising sales budget by 20%. The result was a 15% increase in advertisements. The sales budget may be increased to as much as $50,000/week, but the overall budget for business expenses will not be increased beyond the current level of $200,000/week.

(a) Find the editorial and sales budget figures that maximize profit. Assume that the subscription price remains at $1.50/week, and the advertising price stays at $250/page. Use the five-step method, and model as a constrained optimization problem. Solve using the method of Lagrange multipliers.

(b) Calculate the shadow price for each constraint, and interpret their meaning.

(c) Draw a graph of the feasible region for this problem. Indicate the location of the optimal solution on this graph. Which of the constraints are binding at the optimal solution? How is this related to the shadow prices?

(d) Suppose that cuts in the editorial budget produce an unusually strong negative response in this market. Assume that a 10% cut in the editorial budget causes the paper to lose $p\%$ of its advertising and $2p\%$ of its subscribers. Determine the smallest value of p for which the paper would be better off not to cut the editorial budget.

10. A shipping company has the capacity to move 100 tons/day by air. The company charges $250/ton for air freight. Besides the weight constraint, the company can only move 50,000 ft^3 of cargo per day because of limited volume of aircraft storage compartments. The following cargo are available each day:

Cargo	Weight (tons)	Volume (ft^3/ton)
1	30	550
2	40	800
3	50	400

(a) Determine how much of each cargo should be shipped by air each day in order to maximize revenue. Use the five-step method, and model as a constrained optimization problem. Solve using Lagrange multipliers.

(b) Calculate the shadow prices for each constraint, and interpret their meaning.

(c) The company has the capability to reconfigure some of its older planes to increase the size of the cargo areas. The alterations would cost $200,000 per plane and would add 2,000 ft^3 per plane. The weight limits would stay the same. Assuming that the planes fly 250 days per year and that the remaining lifetime of the older planes is around 5 years, would it be worthwhile to make the alterations? To how many planes?

Further Reading

Beightler, C., Phillips, D. and Wilde, D. (1979). *Foundations of Optimization*, 2nd ed., Prentice–Hall, Englewood Cliffs, New Jersey.

Courant, R. (1937). *Differential and Integral Calculus*, vol. II, Wiley, New York.

Edwards, C. (1973). *Advanced Calculus of Several Variables*, Academic Press, New York.

Hundhausen, J., and Walsh, R. *The Gradient and Some of its Applications*, UMAP module 431.

Hundhausen, J., and Walsh, R. *Unconstrained Optimization*, UMAP module 522.

Nievergelt, Y. *Price Elasticity of Demand: Gambling, Heroin, Marijuana, Whiskey, Prostitution, and Fish*, UMAP module 674.

Nevison, C. *Lagrange Multipliers: Applications to Economics*, UMAP module 270.

Peressini, A. *Lagrange Multipliers and the Design of Multistage Rockets*, UMAP module 517.

Chapter Three

Computational Methods for Optimization

The preceding chapters have discussed some of the analytic techniques for solving optimization problems. These techniques form the basis for most optimization models. In this chapter we will study some of the computational problems that arise in real applications and discuss a few of the most popular methods for dealing with them.

3.1 One-Variable Optimization

Even for simple one-variable optimization problems, the task of locating global extreme points can be exceedingly difficult. Real problems are usually messy. Even when the functions involved are differentiable everywhere, the computation of the derivative is often complicated. The worst part, however, is solving the equation $f'(x) = 0$. The plain and simple fact is that most equations cannot be solved analytically. The best we can do in most instances is to find an approximate solution by graphical or numerical techniques.

Example 3.1 Reconsider the pig problem of Example 1.1, but now take into account the fact that the growth rate of the pig is not constant. Assume that the pig is young, so that the growth rate is increasing. When should we sell the pig for maximum profit?

We will use the five-step method. Step 1 will be to modify the work we did in Section 1.1, as summarized in Fig. 1.1. Now we cannot assume that $w = 200 + 5t$. What would be a reasonable assumption to represent an increasing rate of growth? There are, of course, many possible answers to this question. Let us suppose for now that the growth rate of the pig is proportional to its weight. In other words, let us assume that

$$\frac{dw}{dt} = cw. \tag{1}$$

From the fact that $dw/dt = 5$ lbs/day when $w = 200$ lbs, we conclude that $c = 0.025$. This leaves us with a simple differential equation to solve for w, namely

$$\frac{dw}{dt} = 0.025w, \quad w(0) = 200. \tag{2}$$

We can solve Eq. (2) by separation of variables to obtain

$$w = 200e^{0.025t}. \tag{3}$$

Since all of our other assumptions are unchanged from what was presented in Fig. 1.1, this concludes step 1.

Step 2 is to select our modeling approach. We will model the problem as a one-variable optimization problem. The general solution procedure for one-variable optimization problems was outlined in Section 1.1. In this section we will explore some computational methods that can be used to implement this general solution procedure. Computational methods such as those we present here are often needed in real problems when calculations become either too hard or too tedious to perform by hand.

Step 3 is to formulate the model. The only difference between the present case and the problem formulation of Section 1.1 is that we have to replace the weight equation $w = 200 + 5t$ by Eq. (3). This leads to the new objective function

$$\begin{aligned} y &= f(x) \\ &= (0.65 - 0.01x)(200e^{0.025x}) - 0.45x, \end{aligned} \tag{4}$$

and our problem is to maximize the function in Eq. (4) over the set $S = \{x : x \geq 0\}$.

Step 4 is to solve the model. We will use the graphical method. Good graphing programs for personal computers (IBM PC and compatibles, MacIntosh, etc.) and graphing calculators are widely available. We start our graphical analysis of this problem by producing a graph of the function in Eq. (4) on the same scale as Fig. 1.2, our graph of the original objective function. In this case we are left with the feeling that there is more to see on the graph of this function. We would say that Fig. 3.1 is not a *complete graph* of this function over the set $S = [0, \infty)$. Figure 3.2 *is* a complete graph. It shows all of the important features we need for the solution of our problem. How do we know when we have a complete graph? There is no simple answer to this question. Graphing is an exploratory technique. You need to experiment and to use good judgment. Of course, we do not need to look at negative values of x, but we also need not look beyond $x = 65$. After this point our formula says that the price for pigs is negative, which is clearly nonsense.

From the graph in Fig. 3.2 we can conclude that the maximum occurs around $x = 20$ and $y = f(x) = 140$. To obtain a better estimate, we can zoom in on the maximum point of the graph. Figures 3.3 and 3.4 show the outcome of successive zooms. On the basis of the graph in Fig. 3.4, we would estimate that the maximum occurs at

$$x = 19.5$$
$$y = f(x) = 139.395. \tag{5}$$

At this point we have found the location of the maximum to three significant digits, and the value of the maximum to six significant digits. Since $f'(x) = 0$ at the maximum, the function $f(x)$ is quite insensitive to changes in x near this point, so we are able to obtain more accuracy for $f(x)$ than for x.

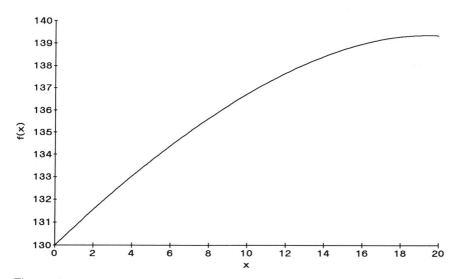

Figure 3.1 Graph of net profit $f(x)$ versus time to sell x for the pig problem with nonlinear weight model.

$$f(x) = (0.65 - 0.01x)(200e^{0.025x}) - 0.45x$$

Step 5 is to answer the question. After taking into account the fact that the growth rate of the young pig is still increasing, we now recommend waiting 19 or 20 days to sell. This should result in a net profit of approximately $140.

The graphical method used in step 4 above to locate the optimum point (see Eq. (5)) did not produce a high degree of accuracy. This is acceptable for the present problem, because we do not need a high degree of accuracy. While it is true that the graphical method can be made to produce higher accuracy (by zooming in on the optimum point over and over again), there are more efficient computational methods that should be used in such cases. We will look at some of these next, in the course of our sensitivity analysis.

Let us examine the sensitivity of the optimum coordinates in Eq. (5) to the growth rate $c = 0.025$ for the young pig. One way to do this would be to repeat our graphical analysis for several different values of the parameter c. However, this would be tedious. We would prefer a more efficient method.

Let us begin by generalizing the model. Now we are assuming that

$$\frac{dw}{dt} = cw, \quad w(0) = 200, \tag{6}$$

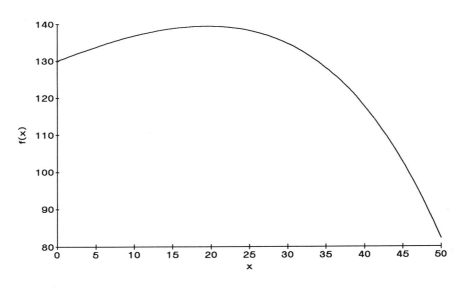

Figure 3.2 Complete graph of net profit $f(x)$ versus time to sell x for the pig problem with nonlinear weight model.

$$f(x) = (0.65 - 0.01x)(200e^{0.025x}) - 0.45x$$

so that

$$w = 200e^{ct}. \tag{7}$$

This leads to the objective function

$$f(x) = (0.65 - 0.01x)(200e^{cx}) - 0.45x. \tag{8}$$

From our graphical analysis, we know that for $c = 0.025$, the optimum occurs at an interior critical point, at which point $f'(x) = 0$. Since f is a continuous function of c, it seems reasonable to conclude that the same holds for values of c near 0.025. In order to locate this interior critical point, we need to compute the derivative $f'(x)$ and solve the equation $f'(x) = 0$. The first part of this process (computing the derivative) is relatively easy. There is a standard method for computing derivatives, which you learned in one-variable calculus. It can be applied to virtually any differentiable function. For complicated expressions the derivative can also be computed using a computer algebra system (MAPLE, MATHEMATICA, DERIVE, etc.) or a hand calculator (such as the HP–48). In our problem, it is not too difficult to compute by hand that

$$f'(x) = 200ce^{cx}(0.65 - 0.01x) - 2e^{cx} - 0.45. \tag{9}$$

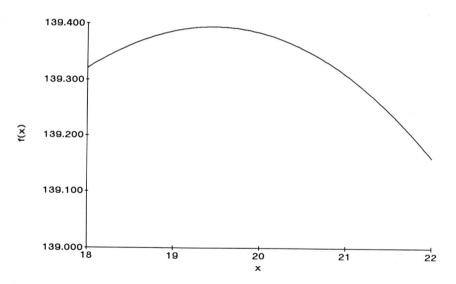

Figure 3.3 First zoom-in on the graph of net profit $f(x)$ versus time to sell x for the pig problem with nonlinear weight model.

$$f(x) = (0.65 - 0.01x)(200e^{0.025x}) - 0.45x$$

The second part of the process is to solve the equation

$$200ce^{cx}(0.65 - 0.01x) - 2ce^{cx} - 0.45 = 0. \tag{10}$$

You can try to solve this equation either by hand, or with a computer algebra system, but you will not succeed in obtaining an algebraic solution. The plain and simple fact is that most equations cannot be solved by algebraic methods. While there does exist a general algebraic method for computing derivatives, there is no general algebraic method for solving equations. Even for polynomials it is known that there can be no generally useful algebraic method of finding roots (i.e., there can be no analogue of the quadratic formula) for degree five or above. This is why it is often necessary to resort to *numerical approximation* methods to solve algebraic equations.

We will use Newton's method to solve Eq. (10). You probably learned about Newton's method in one-variable calculus. We are given a differentiable function $F(x)$ and an approximation x_0 to the root $F(x) = 0$. Newton's method produces a sequence of increasingly accurate estimates x_1, x_2, x_3, ..., to the actual root. Once we are sufficiently close to the root, each successive approximation produced by Newton's method is accurate to twice as many decimal places as the preceding estimate.

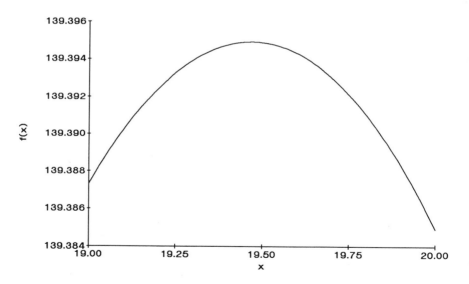

Figure 3.4 Second zoom-in on the graph of net profit $f(x)$ versus time to sell x for the pig problem with nonlinear weight model.

$$f(x) = (0.65 - 0.01x)(200e^{0.025x}) - 0.45x$$

Algorithm: Newton's Method

Variables: $x(n)$ = approximate location of root after n iterations.
N = number of iterations

Input: $x(0)$, N

Process: Begin
for $n = 1$ to N do
  Begin
    $x(n) \longleftarrow x(n-1) - F(x(n-1))/F'(x(n-1))$
  End
End

Output: $x(N)$

Figure 3.5 Pseudocode for Newton's method in one variable.

Figure 3.5 presents Newton's method in a form called *pseudocode*. This is a standard method for describing a numerical algorithm. It is a fairly simple matter to transform pseudocode into a working computer program in any high-level computer language (BASIC, FORTRAN, C, PASCAL, etc.) or to implement on a spreadsheet. It is also possible to program in most computer algebra systems. For the numerical methods in this book, any of these options can be used. It is not recommended that these algorithms be implemented by hand.

In our problem we want to use Newton's method to find a root of the equation

$$F(x) = 200ce^{cx}(0.65 - 0.01x) - 2ce^{cx} - 0.45 = 0. \tag{11}$$

For values of c near 0.025 we expect to find a root near the point $x = 19.5$. We used a computer implementation of Newton's method to produce the results in Table I. For each value of c we performed $N = 10$ iterations starting at the point $x(0) = 19.5$. An additional sensitivity run with $N = 15$ was used to verify the accuracy of our results. In order to relate our sensitivity results back to the original data in the problem, in Figure 3.6 we have plotted the root x, which represents the optimal time to sell, against the growth rate

$$g = 200c, \tag{12}$$

which was originally given as $g = 5$ lbs/day. To obtain a numerical estimate of sensitivity, we solve Eq. (11) once more, setting $c = 0.02525$ (a 1% increase).

Table I.   Sensitivity of best time to sell x to growth rate
parameter c for the pig problem with nonlinear weight model.

c	r
0.022	11.626349
0.023	14.515929
0.024	17.116574
0.025	19.468159
0.026	21.603681
0.027	23.550685
0.028	25.332247

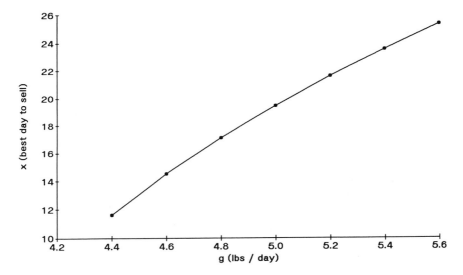

Figure 3.6 Graph of best time to sell x versus growth rate g for the pig problem with
nonlinear weight model.

The solution found was

$$x = 20.021136,$$

which represents a 2.84% increase in x. Thus, we estimate that $S(x, c) = 2.84$.
    Since

$$g = 200c,$$

we can easily show that

$$S(x, g) = S(x, c) = 2.84.$$

Also, if we let h denote the initial weight of the pig (we assumed $h = 200$ lbs), then since

$$h = 5/c,$$

we have

$$
\begin{aligned}
S(x, h) &= \frac{dx}{dh} \cdot \frac{h}{x} \\
&= \left(\frac{dx/dc}{dh/dc} \right) \left(\frac{5/c}{x} \right) \\
&= -S(x, c) = -2.84.
\end{aligned}
$$

In fact, it is generally true that if y is proportional to z, then

$$S(x, y) = S(x, z), \tag{13}$$

and if y is inversely proportional to z, then

$$S(x, y) = -S(x, z). \tag{14}$$

Now we have computed the sensitivity of x to both the initial weight of the pig and the growth rate of the pig. The other sensitivities are unchanged from the original problem considered in Chapter 1, since the other parameters appear in the objective function in the same manner as before.

The fact that our optimal solution in the present case differs so markedly from what we found in Chapter 1 (19 or 20 days here versus 8 days there) raises some important questions about the robustness of our model. There is also serious concern about whether the assumption

$$p = 0.65 - 0.01t$$

would be valid over a three-week period. Alternative models for price could certainly be considered, along with more sophisticated models of the growth process for a representative pig. Some of these issues of robustness are addressed in the exercises at the end of this chapter. What we can say now is this: If the pig's rate of growth does not diminish, and if the price decline does not accelerate, then we should hold onto the pig for another week. At that time, we can reevaluate the situation on the basis of new data.

3.2 Multivariable Optimization

The practical problems associated with locating the global optimum of a function of several variables are similar in many ways to those discussed in the preceding section. Additional complications arise because of the dimension of the problem. Graphical techniques are not available in dimensions $n > 3$, and solving $\nabla f = 0$ becomes more complicated as the number of independent variables increases. Constrained optimization is also more difficult because the geometry of the feasible region can be more complicated.

Example 3.2   A suburban community intends to replace its old fire station with a new facility. The old station was located at the historical city center. City planners intend to locate the new facility more scientifically. A statistical analysis of response-time data yielded an estimate of $3.2 + 1.7r^{0.91}$ minutes required to respond to a call r miles away from the station. (The derivation of this formula is the subject of Exercises 16 and 17 in Chapter 8.) Estimates of the frequency of calls from different areas of the city were obtained from the fire chief. They are presented in Figure 3.7. Each block represents one square mile, and the numbers inside each block represent the number of emergency calls per year for that block. Find the best location for the new facility.

We will represent locations on the city map by coordinates (x, y), where x is the distance in miles to the west side of town and y is the distance in miles to the south side. For example, $(0, 0)$ represents the lower left-hand corner of the map, and $(0, 6)$ represents the upper left-hand corner. For simplicity we will divide the city into nine 2×2-mile squares and assume that each emergency is located at the center of the square. If (x, y) is the location of the new fire station, the average response time to a call is $z = f(x, y)$, where

$$z = 3.2 + 1.7[6\sqrt{(x-1)^2 + (y-5)^2}^{0.91}$$
$$+ 8\sqrt{(x-3)^2 + (y-5)^2}^{0.91} + 8\sqrt{(x-5)^2 + (y-5)^2}^{0.91}$$
$$+ 21\sqrt{(x-1)^2 + (y-3)^2}^{0.91} + 6\sqrt{(x-3)^2 + (y-3)^2}^{0.91}$$
$$+ 3\sqrt{(x-5)^2 + (y-3)^2}^{0.91} + 18\sqrt{(x-1)^2 + (y-1)^2}^{0.91} \tag{15}$$
$$+ 8\sqrt{(x-3)^2 + (y-1)^2}^{0.91} + 6\sqrt{(x-5)^2 + (y-1)^2}^{0.91}]/84.$$

The problem is to minimize $z = f(x, y)$ over the feasible region $0 \leq x \leq 6$, $0 \leq y \leq 6$.

Figure 3.8 shows a graph of the objective function f over the feasible region. It appears as though f attains its minimum at the unique interior point at which

3	0	1	4	2	1
2	1	1	2	3	2
5	3	3	0	1	2
8	5	2	1	0	0
10	6	3	1	3	1
0	2	3	1	1	1

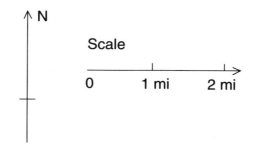

Figure 3.7 Map showing the number of emergency calls per year in each one square mile area of the city.

$\nabla f = 0$. Now it is certainly possible to compute ∇f in this problem, but it is not possible to solve $\nabla f = 0$ algebraically. Further graphical analysis is possible, but this is especially cumbersome for 3-D graphs. What is needed here is a simple global method for estimating the minimum.

Figure 3.9 presents an algorithm for random search. This optimization method simply picks N feasible points at random and selects the one that yields the smallest value of the objective function. The notation Random $\{S\}$ denotes a randomly selected point from the set S. A computer implementation of random search was applied to the function in Eq. (15) with $a = 0$, $b = 6$, $c = 0$, $d = 6$, and $N = 1,000$. The resulting estimate of the minimum occurred at

$$x\ min = 1.66$$
$$y\ min = 2.73 \tag{16}$$
$$z\ min = 6.46.$$

Since this algorithm involves random numbers, a repetition of the same computer

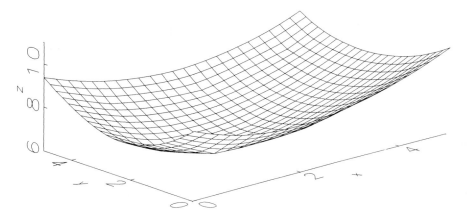

Figure 3.8 Graph of average response time $z = f(x, y)$ versus map location (x, y) for the facility location problem.

$$f(x, y) = 3.2 + 1.7[6\sqrt{(x-1)^2 + (y-5)^2}^{-0.91}$$
$$+ 8\sqrt{(x-3)^2 + (y-5)^2}^{-0.91} + 8\sqrt{(x-5)^2 + (y-5)^2}^{-0.91}$$
$$+ 21\sqrt{(x-1)^2 + (y-3)^2}^{-0.91} + 6\sqrt{(x-3)^2 + (y-3)^2}^{-0.91}$$
$$+ 3\sqrt{(x-5)^2 + (y-3)^2}^{-0.91} + 18\sqrt{(x-1)^2 + (y-1)^2}^{-0.91}$$
$$+ 8\sqrt{(x-3)^2 + (y-1)^2}^{-0.91} + 6\sqrt{(x-5)^2 + (y-1)^2}^{-0.91}]/84$$

Algorithm:	Method of Random Search

Variables:	a	= lower limit on x
	b	= upper limit on y
	c	= lower limit on y
	d	= upper limit on y
	N	= number of iterations
	$x\,min$ = approximate x coordinate of minimum	
	$y\,min$ = approximate y coordinate of minimum	
	$z\,min$ = approximate value of $F(x,\,y)$ at minimum	

Input:	$a,\,b,\,c,\,d,\,N$

Process:	Begin

$x \longleftarrow$ Random $\{[a,\,b]\}$
$y \longleftarrow$ Random $\{[c,\,d]\}$
$z\,min \longleftarrow F(x,\,y)$
for $n = 1$ to N do
   Begin
      $x \longleftarrow$ Random $\{[a,\,b]\}$
      $y \longleftarrow$ Random $\{[c,\,d]\}$
      $z \longleftarrow F(x,\,y)$
      if $z < z\,min$ then
         $x\,min \longleftarrow x$
         $y\,min \longleftarrow y$
         $z\,min \longleftarrow z$
   End
End

Output:	$x\,min,\,y\,min,\,z\,min$

Figure 3.9 Pseudocode for the method of random search.

implementation with the same inputs may produce slightly different outputs. The accuracy of random search is roughly the same as if the N points were to lie on an equally spaced grid over the entire feasible set. Such a grid would contain 32×32 points ($32^2 \approx 1,000$), so we would be accurate to within $6/32 \approx 0.2$ in both x and y. Since $\nabla f = 0$ at the minimum, we obtain much better accuracy in z. An alternative to random search would be a grid search (examine $z = F(x, y)$ at N equally spaced points). The performance of grid search is essentially the same as random search, but random search is more flexible and easier to implement.

The estimates in Eq. (16) of the optimum location (1.7, 2.7) and the resulting average response time of 6.46 minutes were obtained by evaluating the objective function at $N = 1,000$ random points in the feasible region. Better accuracy could be obtained by increasing N. However, the behavior of this simple, global method does not encourage such an approach. Each additional decimal place of accuracy requires increasing N by *a factor of one hundred*. Hence, this method is only suitable for obtaining a rough approximation of the optimum. For the present problem the answer we found is good enough. The simplifying assumptions we made earlier introduced errors on the order of one mile in emergency location, so there is no point in demanding more accuracy now. It is enough to state that the facility should be located around (1.7, 2.7) on the map, for an average response time of around 6.5 minutes. The exact location will in any case depend on several factors not incorporated into our model, such as the location of roads and the availability of land in the area of the optimal site.

It is important to provide an estimate of the sensitivity of response time to the eventual facility location. Since $\nabla f = 0$ at the optimal location, we do not expect f to vary much near (1.7, 2.7). To obtain a more concrete understanding of the sensitivity of f to (x, y) near the optimum, we reran the random search program, replacing f by $-f$ and bounding $1.5 \leq x \leq 2, 2.5 \leq y \leq 3$. After $N = 100$ observations we found that the maximum of f over this region was approximately 6.49 minutes, or about .03 minutes longer than the observed optimum. It does not matter in any practical sense where in this half-mile square area we locate the station.

The random search methods employed in the preceding example are simple but slow. For some problems, where greater accuracy is required, such methods are unsuitable. If the functions involved are as complex as the objective in Example 3.2, then it will be very hard to obtain an accurate answer. More accurate and efficient methods for global optimization of functions of more than one variable are almost always based on the gradient. We now present an example in which these methods are more tractable, because the gradient is easier to compute.

Example 3.3   A manufacturer of lawn furniture makes two types of lawn chairs, one with a wood frame and one with a tubular aluminum frame. The wood-frame model costs $18 per unit to manufacture, and the aluminum-frame model costs $10 per unit. The company operates in a market where the number of units that can be sold depends on the price. It is estimated that in order to sell x units per day of the wood-frame model and y units per day of the aluminum-frame model, the selling price cannot exceed $10 + 31x^{-0.5} + 1.3y^{-0.2}$ $/unit for wood-frame chairs, and $5 + 15y^{-0.4} + 0.8x^{-0.08}$ $/unit for aluminum-frame chairs. Find the optimal production levels.

The objective is to maximize the profit function $z = f(x, y)$ $/day over the feasible set of production levels $x \geq 0$, $y \geq 0$, where

$$z = x(10 + 31x^{-0.5} + 1.3y^{-0.2}) - 18x$$
$$+ y(5 + 15y^{-0.4} + 0.8x^{-0.08}) - 10y. \tag{17}$$

Figure 3.10 shows a graph of f. The graph suggests that the maximum occurs around

$$x = 5$$
$$y = 5$$
$$z = 50$$

at the unique point in the feasible region at which $\nabla f = 0$. We calculate the

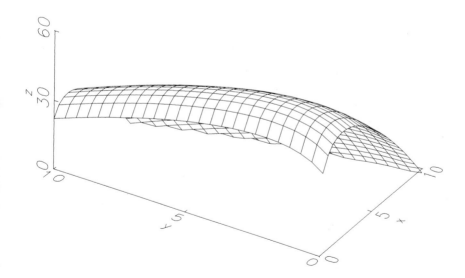

Figure 3.10 Graph of profit $z = f(x, y)$ versus the number x of wood-frame chairs and the number y of aluminum-frame chairs produced per day in the lawn chair problem.

$$f(x, y) = x(10 + 31x^{-0.5} + 1.3y^{-0.2}) - 18x$$
$$+ y(5 + 15y^{-0.4} + 0.8x^{-0.08}) - 10y$$

gradient $\nabla f(x, y) = (\partial z/\partial x, \; \partial z/\partial y)$ to get

$$\frac{\partial z}{\partial x} = 15.5x^{-0.5} - 8 + 1.3y^{-0.2} - 0.064yx^{-1.08}$$

$$\frac{\partial z}{\partial y} = 9y^{-0.4} - 5 + 0.8x^{-0.08} - 0.26xy^{-1.2}.$$

(18)

A random search of $N = 1,000$ points over the region $0 \le x \le 10$, $0 \le y \le 10$ yields an observed maximum of \$52.06 per day at the production level $x = 4.8$ and $y = 5.9$ chairs per day. To obtain a more accurate numerical estimate of the optimum point, we will use a multivariable version of Newton's method to solve the gradient equation $\nabla f = 0$.

The general idea behind Newton's method in one variable is to approximate a nonlinear equation by a linear one. Solving the linear equation yields an approximate solution to the original nonlinear equation. The same idea works for systems of equations. Figure 3.11 gives the algorithm for Newton's method in two variables. We are given two differentiable functions, F and G, and an approximate root $(x_0, \; y_0)$ to the system of equations

$$F(x, y) = 0$$
$$G(x, y) = 0.$$

(19)

Newton's method produces a sequence of increasingly accurate estimates $(x_1, \; y_1)$, $(x_2, \; y_2)$, $(x_3, \; y_3)$, $\ldots$, to the actual root of the system in Eq. (19). Once we are sufficiently close to the root, each successive approximation is accurate to twice as many decimal places as the preceding estimate. Just as in the one-variable case, the sequence of estimates produced by Newton's method converges rapidly to the actual root. For more details on Newton's method in several variables, see Press, et al., (1987), p. 269.

In our problem we have

$$F(x, y) = 15.5x^{-0.5} - 8 + 1.3y^{-0.2} - 0.064yx^{-1.08}$$
$$G(x, y) = 9y^{-0.4} - 5 + 0.8x^{-0.08} - 0.26xy^{-1.2},$$

(20)

and so we can compute that

$$\frac{\partial F}{\partial x} = 0.06912yx^{-2.08} - 7.75x^{-1.5}$$

$$\frac{\partial F}{\partial y} = -0.064x^{-1.08} - 0.26y^{-1.2}$$

$$\frac{\partial G}{\partial x} = -0.064x^{-1.08} - 0.26y^{-1.2}$$

$$\frac{\partial G}{\partial y} = 0.312xy^{-2.2} - 3.6y^{-1.4}.$$

(21)

Algorithm: Newton's Method in Two Variables

Variables: $x(n)$ = approximate x coordinate of the root after n iterations.
$y(n)$ = approximate y coordinate of the root after n iterations.
N = number of iterations

Input: $x(0),\ y(0),\ N$

Process: Begin
                for $n = 1$ to N do
                  Begin
                  $q \longleftarrow \partial F/\partial x(x(n-1),\ y(n-1))$
                  $r \longleftarrow \partial F/\partial y(x(n-1),\ y(n-1))$
                  $s \longleftarrow \partial G/\partial x(x(n-1)\ y(n-1))$
                  $t \longleftarrow \partial G/\partial y(x(n-1),\ y(n-1))$
                  $u \longleftarrow -F(x(n-1),\ y(n-1))$
                  $v \longleftarrow -G(x(n-1),\ y(n-1))$
                  $D \longleftarrow qt - rs$
                  $x(n) \longleftarrow x(n-1) + (ut - vr)/D$
                  $y(n) \longleftarrow y(n-1) + (qv - su)/D$
                  End
                End

Output: $x(N),\ y(N)$

Figure 3.11 Pseudocode for Newton's method in two variables.

We used a computer implementation of Newton's method in two variables starting
at $x(0) = 5,\ y(0) = 5$. After $N = 10$ iterations we obtained

$$x = 4.68959$$
$$y = 5.85199$$
                                                                      (22)

as our estimate of the root. A further sensitivity run with $N = 15$ was made to
confirm these results. Substituting back into Eq. (17) yields $z = 52.0727$. Hence,
the optimal solution to the lawn chair problem is to produce 4.69 wood-frame
chairs and 5.85 aluminum-frame chairs per day, which should result in a net profit
of \$52.07 per day.

Multivariable optimization problems with constraints may be harder to solve.
In some cases the optimum occurs in the interior of the feasible region, so that we
may treat the problem as if there were no constraints. When the optimum occurs

on the boundary, the situation is more complex. There are no simple, generally effective computational algorithms for solving such problems. The methods that do exist are tailored to the particular features of some special class of problems. In the next section we will discuss the most important of these.

3.3 Linear Programming

Multivariable optimization problems with constraints are almost always difficult to solve. A variety of computational techniques have been developed to handle special types of multivariable optimization problems, but good general methods do not yet exist even at the most sophisticated levels. The area of research that considers the development of new computational methods for such problems is called nonlinear programming, and it is very active.

The simplest type of multivariable constrained optimization problem is one where both the objective function and the constraint functions are linear. The study of computational methods for such problems is called *linear programming*. Software packages for linear programming are widely available and are in frequent use for problems in manufacturing, investment, farming, transportation, and government. Typical large-scale problems involve thousands of decision variables and thousands of constraints. There are many well-documented cases where an operations analysis based on a linear programming model has generated savings in the millions of dollars. Details can be found in the literature on operations research and management science.

Example 3.4 A family farm has 625 acres available for planting. The crops the family is considering are corn, wheat, and oats. It is anticipated that 1,000 acre-ft of water will be available for irrigation, and the farmers will be able to devote 300 hours of labor per week. Additional data are presented in Table II. Find the amount of each crop which should be planted for maximum profit.

We will use the five-step method. The results of step 1 are shown in Figure 3.12. Step 2 is to select the modeling approach. We will model this problem as a linear programming problem.

The standard (inequality) form of a linear programming model is as follows: Maximize the objective function $y = f(x_1, \ldots, x_n) = c_1 x_1 + \cdots + c_n x_n$ over the feasible region defined by the constraints

$$a_{11} x_1 + \cdots + a_{1n} x_n \leq b_1$$

$$\vdots \qquad \qquad \vdots \qquad \vdots \qquad \qquad (23)$$

$$a_{m1} x_1 + \cdots + a_{mn} x_n \leq b_m$$

and $x_1 \geq 0, \ldots, x_n \geq 0$. This is a special case of the multivariable con-

Table II. Data for the farm problem of Example 3.4.

Requirements per acre	Crop		
	Corn	Wheat	Oats
Irrigation (acre-ft)	3.0	1.0	1.5
Labor (person-hrs/week)	0.8	0.2	0.3
Yield ($/acre)	400	200	250

strained optimization problem discussed in Chapter 2. Let

$$g_1(x_1, \ldots, x_n) = a_{11}x_1 + \cdots + a_{1n}x_n$$
$$\vdots$$
$$g_m(x_1, \ldots, x_n) = a_{m1}x_1 + \cdots + a_{mn}x_n$$

(24)

and

$$g_{m+1}(x_1, \ldots, x_n) = x_1$$
$$\vdots$$
$$g_{m+n}(x_1, \ldots, x_n) = x_n.$$

The constraints can be written in the form $g_1 \leq b_1, \ldots, g_m \leq b_m$, and $g_{m+1} \geq 0, \ldots, g_{m+n} \geq 0$. The set of $(x_1, \ldots, x_n)$ which satisfies these constraints is called the *feasible region*. It represents all feasible values for the decision variables $x_1, \ldots, x_n$. Since $\nabla f = (c_1, \ldots, c_n)$ can never be zero, the function cannot attain its maximum in the interior of the feasible region. At a maximum on the boundary we must have

$$\nabla f = \lambda_1 \nabla g_1 + \cdots + \lambda_{m+n} \nabla g_{m+n}$$

(25)

with $\lambda_i \neq 0$ only if the ith constraint is binding. For $i = 1, \ldots, m$, the value of the Lagrange multiplier λ_i represents the potential increase in the maximum value of the objective function $y = f(x_1, \ldots, x_n)$ that would result from relaxing the ith constraint by one unit (changing it to $g_i(x_1, \ldots, x_n) \leq b_i + 1$). Computation of the optimal solution to a linear programming problem is usually obtained by computer, using a variation of the *simplex method*. This method is based on the fact (which you can easily verify) that the optimal solution must occur at one of the corner points of the feasible region. Rather than go into the details involved in the algebra of the simplex method, we will

Variables: $x_1 = $ acres of corn planted

$x_2 = $ acres of wheat planted

$x_3 = $ acres of oats planted

$w \ = $ irrigation required (acre-ft)

$l \ \ = $ labor required (person-hrs/wk)

$t \ \ = $ total acreage planted

$y \ \ = $ total yield ($)

Assumptions: $w = 3.0x_1 + 1.0x_2 + 1.5x_3$

$l \ = 0.8x_1 + 0.2x_2 + 0.3x_3$

$t \ = x_1 + x_2 + x_3$

$y \ = 400x_1 + 200x_2 + 250x_3$

$w \leq 1,000$

$l \ \leq 300$

$t \ \leq 625$

$x_1 \geq 0; \ x_2 \geq 0; \ x_3 \geq 0$

Objective: Maximize y

Figure 3.12 Results of step 1 of the farm problem.

concentrate on what you need to know in order to use a linear programming package correctly.

In the simplex method, the coordinates of a corner point are computed using the alternative (equality) form of a linear programming problem: Maximize $y = c_1 x_1 + \cdots + c_n x_n$ over the set

$$
\begin{aligned}
a_{11}x_1 + \cdots + a_{1n}x_n + x_{n+1} \quad\quad\quad\quad &= b_1 \\
a_{21}x_1 + \cdots + a_{2n}x_n \quad\quad\ + x_{n+2} \quad\quad &= b_2 \\
\vdots \quad\quad\quad\quad\quad\quad & \\
a_{m1}x_1 + \cdots + a_{mn}x_n \quad\quad\quad\quad + x_{n+m} &= b_m
\end{aligned}
\tag{26}
$$

and $x_1 \geq 0, \ldots, x_{n+m} \geq 0$. The variable x_{n+i} is called a *slack variable*, because it represents the amount of slack remaining in the ith constraint. The *i*th constraint is binding when the slack variable $x_{n+i} = 0$. The coordinates of a corner point can be obtained by setting n of the variables $x_1, \ldots, x_{m+n}$ equal to zero, and then solving the resulting m equations in m unknowns. The corner point is feasible if the remaining m variables (called *basic variables*) turn out to be nonnegative.

Suppose that we have a moderate size linear programming problem, with $n = 50$ variables and $m = 100$ constraints. The number of corner points is equal to the number of possible choices for which 50 of the 150 variables (50 decision variables plus 100 slack variables) we set equal to zero. The number of possible ways to choose 50 out of 150 is $(150!)/(50!)(100!)$, or around 2×10^{40}. A computer program that could check one corner point every nanosecond would take around 8×10^{30} years to solve this problem, which is representative of a typical linear programming application. The simplex method calculates only a selected subset of the corner points (a very small fraction of the total). A linear program of size $n = 50$, $m = 100$ can be solved quickly on a mainframe computer. As of this writing, a problem of this size is at the upper limits of what can be solved in a reasonable amount of time on a personal computer, but technological advances will certainly push these limits further in the near future. As a general rule, the execution time for a linear program using the simplex method is proportional to m^3, so that an order-of-magnitude improvement in processor speed more than doubles the size of the linear program that can be handled by a given machine. For the problems in this book, any good computer implementation of the simplex method will be adequate. It is not recommended that these problems be solved by hand.

Step 3 is to formulate the linear programming model in the standard form. In our problem the decision variables are the number of acres of each crop x_1, x_2, and x_3. We want to maximize the total yield, $y = 400x_1 + 200x_2 + 250x_3$, over the set

$$3.0x_1 + 1.0x_2 + 1.5x_3 \leq 1,000$$
$$0.8x_1 + 0.2x_2 + 0.3x_3 \leq 300 \qquad (27)$$
$$x_1 + x_2 + x_3 \leq 625$$

and $x_1 \geq 0, x_2 \geq 0, x_3 \geq 0$.

Step 4 is to solve the problem. We used a computer implementation of the simplex method called LINDO, written by Linus Schrage. The results of step 4 are shown in Figure 3.13. The optimal solution is $Z = 162,500$ at $x_1 = 187.5$, $x_2 = 437.5$, $x_3 = 0$. Since the slack variables for rows 2 and 4 are zero, the first and third constraints are binding. Since the slack variable for row 3 is equal to 62.5, the second constraint is not binding.

Step 5 is to answer the question. The question was how much of each crop to plant. The optimal solution is to plant 187.5 acres of corn, 437.5 acres of wheat, and no oats. This should yield $162,500. The optimal crop mixture we found uses

```
MAX      400 X1 + 200 X2 + 250 X3
SUBJECT TO
        2)    3 X1 + X2 + 1.5 X3 <=    1000
        3)    0.8 X1 + 0.2 X2 + 0.3 X3 <=    300
        4)    X1 + X2 + X3 <=    625
END
```

```
LP OPTIMUM FOUND AT STEP        2

           OBJECTIVE FUNCTION VALUE

        1)      162500.000

VARIABLE            VALUE            REDUCED COST
        X1       187.500000            .000000
        X2       437.500000            .000000
        X3          .000000           -.000015


        ROW    SLACK OR SURPLUS      DUAL PRICES
        2)          .000000          100.000000
        3)        62.500000             .000000
        4)          .000000           99.999980

NO. ITERATIONS=        2
```

Figure 3.13 Optimal solution to the farm problem using the linear programming package LINDO.

all 625 acres and all 1,000 acre-ft of irrigation water, but only 237.5 of the available 300 person-hours of labor per week. Thus there will be 62.5 person-hours per week that may be devoted to other profitable activities, or to leisure.

We will begin our sensitivity analysis by considering the amount of water available for irrigation. This amount will vary as a result of rainfall and temperature, which determine the status of the farm's irrigation pond. It would also be possible to purchase additional irrigation water from a nearby farm. Figure 3.14 illustrates the effect of one additional acre-ft of irrigation water on our optimal solution. Now we can plant an additional half-acre of corn (a more profitable crop), and in fact we save a little bit of labor (0.3 person-hours per week). The net result is an additional $100 in yield.

```
MAX        400 X1 + 200 X2 + 250 X3
SUBJECT TO
        2)    3 X1 + X2 + 1.5 X3 <=    1001
        3)    0.8 X1 + 0.2 X2 + 0.3 X3 <=    300
        4)    X1 + X2 + X3 <=    625
END


LP OPTIMUM FOUND AT STEP        0

          OBJECTIVE FUNCTION VALUE

     1)      162600.000

VARIABLE          VALUE          REDUCED COST
      X1       188.000000          .000000
      X2       437.000000          .000000
      X3         .000000         -.000015


   ROW    SLACK OR SURPLUS      DUAL PRICES
    2)           .000000        100.000000
    3)         62.200000          .000000
    4)           .000000         99.999980

NO. ITERATIONS=        0
```

Figure 3.14 Optimal solution to the farm problem with one extra acre-foot of water using the linear programming package LINDO.

The $100 is the shadow price for this resource (irrigation water). The farm should be willing to purchase additional irrigation water for up to $100 per acre-ft. Alternatively, it should be unwilling to part with its own irrigation water for less than $100 per acre-ft. In Fig. 3.13, the shadow prices for the three resources (water, labor, and land) are called *dual prices*. They appear next to the corresponding slack variables. An additional acre of land would also be worth $100. Additional labor is worth $0, since there is an excess.

The dollar amount per acre which each crop will yield varies with the weather and the market. Figure 3.15 shows the effect of a slightly higher yield for corn. This would not affect our decision variables x_1, x_2, and x_3 (the amount of each crop we should plant). Our yield increases, of course, by $50x_1 = \$9,375$. It

```
MAX       450 X1 + 200 X2 + 250 X3
SUBJECT TO
       2)    3 X1 + X2 + 1.5 X3 <=    1000
       3)    0.8 X1 + 0.2 X2 + 0.3 X3 <=    300
       4)    X1 + X2 + X3 <=    625
END


LP OPTIMUM FOUND AT STEP        0

          OBJECTIVE FUNCTION VALUE

       1)      171875.000

     VARIABLE          VALUE            REDUCED COST
        X1          187.500000            .000000
        X2          437.500000            .000000
        X3            .000000           12.500000


        ROW    SLACK OR SURPLUS     DUAL PRICES
        2)          .000000         125.000000
        3)        62.500000            .000000
        4)          .000000          75.000000

NO. ITERATIONS=         0
```

Figure 3.15 Optimal solution to the farm problem with a higher yield for corn using the linear programming package LINDO.

is also interesting to note the change in the shadow prices. Water is at more of a premium when corn is more valuable. (Although both the water and the land constraints are binding, it is the water constraint that keeps us from planting more corn in place of wheat.)

Figure 3.16 shows what happens if oats yield a bit more than expected. A very small change in this parameter has a very significant effect on our optimal decision. Now we plant oats instead of wheat. We also plant considerably less corn than before. Apparently our model is quite sensitive to this parameter. This being the case, it seems appropriate to consider the sensitivity to this parameter in more depth.

```
MAX        400 X1 + 200 X2 + 260 X3
SUBJECT TO
        2)    3 X1 + X2 + 1.5 X3 <=    1000
        3)    0.8 X1 + 0.2 X2 + 0.3 X3 <=    300
        4)    X1 + X2 + X3 <=    625
END


LP OPTIMUM FOUND AT STEP        1

        OBJECTIVE FUNCTION VALUE

    1)     168333.300

VARIABLE          VALUE          REDUCED COST
    X1        41.666670            .000000
    X2          .000000          13.333340
    X3       583.333300            .000000


ROW     SLACK OR SURPLUS      DUAL PRICES
    2)          .000000          93.333340
    3)        91.666660            .000000
    4)          .000000         120.000000

NO. ITERATIONS=        1
```

Figure 3.16 Optimal solution to the farm problem with a higher yield for oats using the linear programming package LINDO.

Let c denote the yield ($/acre) for oats, so that our objective function $f(x) = 400x_1 + 200x_2 + cx_3$. Note that the value of c does not affect the shape of the feasible region S. Several additional model runs were made varying c. For $c \leq 250$ the optimal solution is at the corner point $(187.5, 437.5, 0)$, and for $c > 250$ the optimal solution is at the adjacent corner point $(41.6\bar{6}, 0, 583.3\bar{3})$. Both points lie on the line formed by the intersection of the two planes

$$3.0x_1 + 1.0x_2 + 1.5x_3 = 1,000$$
$$x_1 + x_2 + x_3 = 625. \tag{28}$$

Consider the gradient vector $\nabla f = (400, 200, c)$ at any point along this line.

```
MAX       400 X1 + 200 X2 + 250 X3
SUBJECT TO
        2)    2.5 X1 + X2 + 1.5 X3 <=    1000
        3)    0.8 X1 + 0.2 X2 + 0.3 X3 <=    300
        4)    X1 + X2 + X3 <=    625
END


LP OPTIMUM FOUND AT STEP        1

        OBJECTIVE FUNCTION VALUE

    1)       175000.000

VARIABLE            VALUE           REDUCED COST
    X1          250.000000             .000000
    X2          375.000000             .000000
    X3            .000000           16.666680


    ROW    SLACK OR SURPLUS         DUAL PRICES
    2)          .000000           133.333300
    3)        25.000000              .000000
    4)          .000000            66.666660

NO. ITERATIONS=        1
```

Figure 3.17 Optimal solution to the farm problem with a lower water requirement for corn using the linear programming package LINDO.

For $c < 250$ the gradient vector points toward the corner point with $x_3 = 0$. For $c > 250$ the gradient vector points toward the corner point with $x_2 = 0$. As c increases, the gradient vector turns away from the former and toward the latter. At $c = 250$ the gradient vector ∇f is perpendicular to the line through these two points. For this value of c, any point along the line segment joining the two corner points will be optimal.

The practical ramification of our model's sensitivity to the parameter c is that we don't know whether we should plant oats or wheat. A small variation in yield would change our optimal decision. In light of the fact that the $/acre yield varies considerably with the weather and the market, it might be best to present the

```
MAX        400 X1 + 200 X2 + 250 X3 + 200 X4
SUBJECT TO
        2)    3 X1 + X2 + 1.5 X3 + 1.5 X4 <=     1000
        3)    0.8 X1 + 0.2 X2 + 0.3 X3 + 0.25 X4 <=     300
        4)    X1 + X2 + X3 + X4 <=     625
END


LP OPTIMUM FOUND AT STEP        1

           OBJECTIVE FUNCTION VALUE

        1)      162500.000

VARIABLE            VALUE            REDUCED COST
      X1         187.500000             .000000
      X2         437.500000             .000000
      X3           .000000            -.000015
      X4           .000000           49.999980


     ROW    SLACK OR SURPLUS         DUAL PRICES
      2)           .000000         100.000000
      3)         62.500000            .000000
      4)           .000000          99.999980

NO. ITERATIONS=        1
```

Figure 3.18 Optimal solution to the farm problem with the addition of a new crop, barley, using the linear programming package LINDO.

farmer with more than one option. Any crop mixture of the form ($0 \le t \le 1$):

$$x_1 = 187.5t + 41.6\bar{6}(1 - t)$$
$$x_2 = 437.5t + 0(1 - t) \qquad\qquad (29)$$
$$x_3 = 0t + 583.3\bar{3}(1 - t)$$

will use all the available land and irrigation water. There is too much uncertainty in the figures for the \$/acre yield to tell which option would produce the most profit.

Sometimes sensitivity analysis is performed in response to client feedback after the results of an initial study have been presented. Suppose that after the farmer has seen the results of our analysis, a new seed catalog arrives with an advertisement for a new variety of corn. This variety of corn is more expensive, but is supposed to require less irrigation. Figure 3.17 shows the results of a sensitivity run in which we assumed that the corn planted requires only 2.5 acre-feet of irrigation water per acre (instead of 3.0). The new seed corn yields an additional $12,500, and in this case we will of course plant more corn than before. It is also interesting to note that in this case the shadow price for water has increased 33%.

Finally, suppose that the farmer wishes to investigate the addition of a new crop, barley. An acre of barley requires 1.5 acre-ft of water, 0.25 person-hours of labor, and is expected to yield $200. We represent the new crop in our model by adding a decision variable $x_4 =$ acres of barley. Figure 3.18 shows our model run. The results are essentially unchanged from our base case. A mixture of corn and wheat remains our optimal solution, and it is easy to see why. Although barley and wheat yield the same, barley requires more water and more labor.

3.4 Exercises

1. Reconsider the pig problem of Example 1.1, but now suppose that the price for pigs after t days is $p = 0.65e^{-(.01/.65)t}$ cents/pound.
    (a) Show that the price for pigs is falling by 1 cent/day at time $t = 0$. What happens as t increases?
    (b) Find the optimal time to sell the pig. Use the five-step method, and model as a one-variable optimization problem.
    (c) The parameter .01 represents the rate at which price is falling at time $t = 0$. Perform a sensitivity analysis on this parameter. Consider both the best time to sell and the resulting net profit.
    (d) Compare the results of (b) to what was done in Section 1.1, and comment on the robustness of our model.
2. Reconsider the pig problem of Example 1.1, but now suppose that the weight of the pig after t days is $w = 800/(1 + 3e^{-t/30})$ lbs.
    (a) Show that the pig is gaining about 5 lbs/day at $t = 0$. What happens as t increases?
    (b) Find the optimal time to sell the pig. Use the five-step approach, and model as a one-variable optimization problem.
    (c) The parameter 800 represents the eventual mature weight of the pig. Perform a sensitivity analysis for this parameter. Consider both the best time to sell and the profit obtained.
    (d) Compare the results obtained in (b) to what was done in Sections 1.1 and 3.1. Comment on the robustness of our model. What general conclusions can we draw?

3. A statistical algorithm to determine the difference in effectiveness between two alternative treatments requires maximizing the quantity

$$\sum_{(k_1,k_2)\in E} \binom{n_1}{k_1} p_1^{k_1} (1-p_1)^{n_1-k_1} \binom{n_2}{k_2} p_2^{k_2} (1-p_2)^{n_2-k_2}$$

over the set

$$S = \{(p_1,\ p_2) : p_1 - p_2 = \Delta;\ p_1, p_2 \in [0,\ 1]\}.$$

Here E is a subset of the set

$$E_0 = \{(k_1,\ k_2) : k_1 = 0,\ 1,\ 2,\ \ldots,\ n_1;\ k_2 = 0,\ 1,\ 2,\ \ldots,\ n_2\}$$

and $\Delta \in [-1,\ 1]$. Find the maximum in the case $\Delta = -0.1$,

$$E = \{(0,\ 4),\ (0,\ 3),\ (0,\ 2),\ (0,\ 1),\ (1,\ 4),\ (1,\ 3),\ (2,\ 4)\}$$

(Santner, T., et al., (1980).)

4. One method of evaluating the effectiveness of institutional trauma and burn medicine involves maximizing the function

$$f(p_1,\ \ldots,\ p_n) = \frac{\left(A - \sum_{i=1}^{n} p_i\right)}{\sqrt{B + \sum_{i=1}^{n} p_i(1-p_i)}}$$

over the set

$$\{(p_i,\ \ldots,\ p_n) : a_i \le p_i \le b_i \text{ for all } i = 1,\ \ldots,\ n\}.$$

Maximize f in the case $n = 2$, $A = -5.92$, $B = 1.58$, $a_1 = 0.01$, $b_1 = 0.33$, $a_2 = 0.75$, and $b_2 = 0.85$. (Falk, J. et al., (1992))

5. Reconsider the competing species model of Exercise 3, Chapter 2. Assume that a level of effort E boat-days will result in the annual harvest of qEx blue whales and qEy fin whales, where the parameter q (catchability) is assumed to equal approximately 10^{-5}. Given a constant level of effort, assume that population levels will stabilize at the point where growth rate equals harvest rate.

    (a) Assuming that the cost of a whaling expedition is $250 per boat-day, find the level of effort that will maximize profit for the industry in the long run. Use the five-step method, and model as a one-variable optimization problem.

(b) Examine the sensitivity to catchability q. Consider profit, level of effort, and the eventual stable population levels of the whales.

(c) Increasing technology will certainly raise whale catchability. What will be the long-term effects on the whale populations and on the whaling industry?

6. Reconsider the facility location problem of Example 3.2, but now assume that the response time from point (x_0, y_0) to point (x_1, y_1) is proportional to the road travel distance $|x_1 - x_0| + |y_1 - y_0|$.

(a) Find the location that minimizes average response time. Use the five-step method, and model as a multivariable unconstrained optimization problem.

(b) Examine the sensitivity of the optimal location to the estimated number of emergencies in each 2×2-mile sector. Can you draw any general conclusions?

(c) Comment on the robustness of this model. Compare the optimal location to that obtained in the analysis of Section 3.2. What do you think would happen if we assumed that response time was proportional to the straight-line distance $r = \sqrt{(x_1 - x_0)^2 + (y_1 - y_0)^2}$?

7. Reconsider the color TV problem of Example 2.1, but now use numerical methods instead of the analytic methods we employed in Chapter 2.

(a) Determine the production levels x_1 and x_2 that maximize the objective function $y = f(x_1, x_2)$ in Eq. (2) of Chapter 2. Use the two-variable version of Newton's method.

(b) As in Section 2.1, let a denote the price elasticity for 19-inch sets. In part (a) we assumed $a = 0.01$. Now assume that a increases by 10% to $a = .011$ and repeat the optimization problem in part (a). Use your results to obtain a numerical estimate of the sensitivities $S(x_1, a)$, $S(x_2, a)$, and $S(y, a)$. Compare to the results obtained analytically in Section 2.1.

(c) Let b denote the price elasticity for 21-inch sets. Currently $b = 0.01$. As in part (b), use numerical methods to estimate the sensitivities of x_1, x_2, and y to the parameter b.

(d) Compare the analytic methods of Section 2.1 to the numerical methods employed in this problem. Which do you prefer? Explain.

8. Reconsider Exercise 6 in Chapter 2, but now suppose that management has been persuaded to lift the ceiling on advertising expenditures. The assumption that sales vary as a linear function of the advertising budget is probably not reasonable over the wider range of advertising budget figures we now wish to consider. Suppose instead that sales are increased by 1,000 units each time the advertising budget is doubled.

(a) Find the price and the advertising budget that will maximize profit. Use the five-step method, and model as an unconstrained optimization problem.

(b) Determine the sensitivity of the decision variables (price and advertising budget) to price elasticity—the 50% number.

(c) Determine the sensitivity of the decision variables to the advertising agency's estimate of 1,000 new sales each time the advertising budget is doubled.

(d) What goes wrong in part (a) if we assume a linear relationship between advertising budget and sales? Why wasn't this a problem in Exercise 6 of Chapter 2?

9. (Continuation of Exercise 8) Repeat Exercise 8, but now assume an alternative model of the relationship between advertising expenditures and sales. Suppose that doubling the advertising budget results in 1,000 additional sales, but doubling it again results in only 500 additional sales, and so forth. Repeat parts (a) through (c) of Exercise 8. In part (c), determine the sensitivity to the assumption of 1,000 additional sales the first time the advertising budget is doubled. Compare your results to those obtained in Exercise 8, and comment on the robustness of the model.

10. Reconsider the newspaper problem of Exercise 7 in Chapter 2, but now suppose that we choose to maximize our profit margin (profit as a percentage of revenue). Assume that our business expenses remain fixed at $200,000 per week.

(a) Find the subscription price and the advertising price that maximize profit margin. Use the five-step method, and model as an unconstrained optimization problem. Find an approximate solution by the method of random search.

(b) Let $z = f(x, y)$ denote the objective function you obtained in part (a). Use a computer algebra system to determine $F = \partial f/\partial x$, and $G = \partial f/\partial y$. Then determine $\partial F/\partial x$, $\partial F/\partial y$, $\partial G/\partial x$, and $\partial G/\partial y$.

(c) Use Newton's method in two variables to obtain a precise answer to the question in part (a). Use the approximate solution from part (a) as your initial estimate. The required derivatives were calculated in part (b).

(d) If you have not previously solved part (a) of Exercise 7, Chapter 2 then do so now. Use any method. Compare to the results of part (c) above. Does it matter whether we choose to maximize profit or profit margin? Explain.

11. Reconsider the lawn chair problem of Example 3.3. Notice that the objective function $f(x, y)$ tends to infinity as x or y approaches zero, and that $f(x, y)$ is undefined on the lines $x = 0$ and $y = 0$ which form the boundary of the

feasible region. Presumably the estimates of price elasticity are not accurate when extrapolated all the way to $x = 0$ or $y = 0$.

(a) Correct this model deficiency by altering the feasible region.

(b) Comment on the robustness of the decisions you made in part (a).

(c) Show that, for your corrected model, the optimal solution lies in the interior of the feasible region. Locate any local maximum of $f(x, y)$ on the boundary, and show that at every such point ∇f points into the interior.

12. Reconsider the newspaper problem of Exercise 9 in Chapter 2. Solve as a linear programming problem using a computer. Answer questions (a), (b), and (c) from the original problem.

13. Reconsider the cargo problem (Exercise 10 of Chapter 2). Solve as a linear programming problem using a computer. Answer questions (a), (b), and (c) from the original problem.

14. Reconsider the color TV problem of Example 2.2, but make the simplifying assumption that the company makes a profit of $80 per 19-inch set and $100 per 21-inch set.

(a) Find the optimal production levels. Use the five-step method, and solve as a linear programming problem using a computer.

(b) Determine the shadow prices for each constraint, and explain what they mean. Which constraints are binding on the optimal solution?

(c) Determine the sensitivity to the objective function coefficients (profit per set). Consider both profit and optimal production levels.

(d) Draw a graph of the feasible region (see Fig. 2.9) and include a picture of ∇f at the optimum. Describe geometrically what happens to the vector ∇f as we change one of the objective function coefficients. Use this geometric idea to determine how much each objective function coefficient can change before the current optimal solution is no longer optimal.

15. The Burningham textiles company has three textile mills located in the southern U.S. and four distribution centers located in Michigan, New York, California, and Georgia. The estimated yearly output from each mill, the allocation to each warehouse, and the shipping costs are tabulated in Table III.

(a) Find the transportation plan that minimizes shipping costs. Use the five-step method, and solve as a linear programming problem using a computer.

(b) Determine the shadow prices for each of the output constraints. Would it be beneficial to shift production capacity from one mill to another? How much should the company be willing to spend to facilitate the shift?

Table III. Data for the transportation problem of Exercise 15.

		\multicolumn{4}{c}{Shipping cost per truck load (\$)}	
		\multicolumn{4}{c}{Distribution center}	
		MI	NY	CA	GA	Output
Textile	1	430	550	680	700	105
Mill	2	510	590	890	685	160
	3	395	425	910	450	85
Allocation		70	100	105	75	

Further Reading

Beltrami, E. (1977). *Models for Public Systems Analysis*, Academic Press, New York.

Dantzig, G. (1963). *Linear Programming and Extensions*, Princeton University Press, Princeton, New Jersey.

Falk, J. et al. (1992). Bounds on the Trauma Outcome Function via Optimization *Operations Research* 40, Supp. No. 1, S86–S95.

Gearhart, W. and Pierce, J. *Fire Control and Land Management in the Chaparral*, UMAP module 687.

Hillier, F. and Lieberman, J. (1990). *Introduction to Operations Research*, McGraw–Hill, New York.

Maynard, J. *A Linear Programming Model for Scheduling Prison Guards*, UMAP module 272.

Press, W., Flannery, B., Teukolsky, S. and Vetterling, W. (1987). *Numerical Recipies*, Cambridge University Press, New York.

Polack, E. (1971). *Computational Methods in Optimization*, Academic Press, New York.

Santner, T. and Snell, M. (1980). Small-sample confidence intervals for $\rho_1 - \rho_2$ and ρ_1/ρ_2 in 2×2 contingency tables, *Journal of the American Statistical Association* 75, 386–394.

Straffin, P. *Newton's Method and Fractal Patterns*, UMAP module 716.

II

Dynamic Models

Many problems of practical interest involve processes that evolve over time. Dynamic models are used to represent the changing behavior of these systems. Space flight, electrical circuits, chemical reactions, population growth, investments and annuities, military battles, the spread of disease, and pollution control are just a few of the many areas in which extensive use is made of dynamic models.

The five-step method and the fundamental principles of sensitivity analysis and robustness are as relevant and useful for dynamic models as they are for optimization models. We will continue to rely on them as we explore some of the most popular and generally applicable dynamic modeling techniques. In the course of this study, we will also introduce the important modeling concepts of state space, equilibrium, and stability. All of this will be very useful in the third and final section of this book, where we explore stochastic models.

Chapter Four

Introduction to Dynamic Models

As a general rule, dynamic models are easy to formulate and hard to solve. Exact analytic solutions are available only for a few special cases, such as linear systems. Numerical methods usually do not provide a good qualitative understanding of system behavior. Therefore, the application of graphical techniques is usually employed as at least one part of the analysis of dynamic models. Because of the inherent simplicity of graphical techniques, along with their geometrical nature, this chapter also provides us with an ideal opportunity to introduce some of the deepest and most fundamental modeling concepts used for dynamic systems.

4.1 Steady-State Analysis

In this section we will consider the simplest type of dynamic model. The mathematics required are elementary indeed. Even so, the practical applications for this model are numerous, and the absence of too much sophisticated technique leaves us free to concentrate on some of the most fundamental ideas of dynamic modeling.

Example 4.1 In an unmanaged tract of forest area, hardwood and softwood trees compete for the available land and water. The more desirable hardwood trees grow more slowly, but are more durable and produce more valuable timber. Softwood trees compete with the hardwoods by growing rapidly and consuming the available water and soil nutrients. Hardwoods compete by growing taller than the softwoods can and shading new seedlings. They are also more resistant to disease. Can these two types of trees coexist on one tract of forest land indefinitely, or will one type of tree drive the other to extinction?

We will use the five-step method. Let H and S denote the populations of hardwood and softwood trees, respectively. A convenient unit often used by biologists is the biomass (tons per acre of living tree). We need to make some assumptions about the dynamics of these two populations. To begin with, we want to make assumptions that are as simple as possible without neglecting the most fundamental aspects of the problem. Later on we can improve or enrich our model if necessary. It is reasonable to assume that in conditions of unrestricted growth (plenty of room, sunshine, water, and soil nutrients) the growth rate of a species is roughly proportional to the size of the species. Twice as many trees give rise to twice as many little trees. As population increases, members of the same species must compete for resources, and this inhibits growth. Thus, it is reasonable to assume that growth rate is roughly linear in population size for small populations and then falls off as population increases. The simplest growth rate function with these properties is

$$g(P) = rP - aP^2.$$

Here r is the intrinsic growth rate, and $a << r$ is a measure of the strength of resource limitations. If a is smaller, there is more room to grow.

The effect of competition is also due to resource limitations. The presence of hardwood trees limits the amount of sunlight, water, etc., available for the softwoods, and vice versa. The loss in growth rate due to competition depends on the size of both populations. A simple assumption is that this loss is proportional to the product of the two. Given these assumptions about growth and competition, we wish to know whether we can expect one species to die out over time. Figure 4.1 summarizes the results of step 1.

Step 2 is to select the modeling approach. We will model this problem as a dynamic model in steady-state.

We are given functions

$$f_1(x_1, \ldots, x_n)$$

$$\vdots$$

$$f_n(x_1, \ldots, x_n)$$

defined on a subset S of $\mathbb{R}^n$. The functions $f_1, \ldots, f_n$ represent the rate of change of each variable $x_1, \ldots, x_n$ respectively. A point $(x_1, \ldots, x_n)$ in the

Variables: H = hardwood population (tons/acre)
S = softwood population (tons/acre)
g_H = growth rate for hardwoods (tons/acre/year)
g_S = growth rate for softwoods (tons/acre/year)
c_H = loss due to competition for hardwoods (tons/acre/year)
c_S = loss due to competition for softwoods (tons/acre/year)

Assumptions: $g_H = r_1 H - a_1 H^2$
$g_S = r_2 S - a_2 S^2$
$c_H = b_1 S H$
$c_S = b_2 S H$
$H \geq 0, S \geq 0$
$r_1, r_2, a_1, a_2, b_1, b_2$ are positive reals

Objective: Determine whether $H \rightarrow 0$ or $S \rightarrow 0$

Figure 4.1 Results of step 1 for the tree problem.

set S is called an *equilibrium point* provided that

$$f_1(x_1, \ldots, x_n) = 0$$

$$\vdots \qquad\qquad (1)$$

$$f_n(x_1, \ldots, x_n) = 0$$

at this point. The rate of change of each of the variables $x_1, \ldots, x_n$ is then equal to zero, and so the system is at rest.

The variables $x_1, \ldots, x_n$ are called *state variables*, and S is called the *state space*. Since the functions $f_1, \ldots, f_n$ depend only on the current state $(x_1, \ldots, x_n)$ of the system, knowledge of the current state suffices to determine the entire future of the system. What happened in the past does not matter. We only need to know where we are now, not how we got here. When we are at an equilibrium point, defined by Eq. (1), we say that the system is in *steady state*. At this point all of the rates of change are equal to zero. All of the forces acting on the system are in balance. For this reason the equations in (1) are sometimes referred to as the *balance equations*. When a dynamic system is in steady-state, it remains there forever. Since all of the rates of change are equal to zero, any future time will find us in exactly the same place we are right now.

In order to find the equilibrium states of a dynamic system, we need to solve the n equations in n unknowns given by Eq. (1). In very easy cases we can solve by hand. Sometimes we can solve using a computer algebra system. All of the problems in this chapter, including the exercises, can be solved using these techniques. Of course, many real problems give rise to systems of equations that cannot be solved analytically. We will treat such problems in Chapter 6, where we discuss computational methods for dynamical systems. (Alternatively, we could use the multivariable version of Newton's method introduced in Chapter 3.)

Step 3 is to formulate the model. Let $x_1 = H$ and $x_2 = S$ denote our two state variables, defined on the state space

$$\{(x_1, x_2) : x_1 \geq 0, x_2 \geq 0\}.$$

The steady-state equations are

$$r_1 x_1 - a_1 x_1^2 - b_1 x_1 x_2 = 0$$
$$r_2 x_2 - a_2 x_2^2 - b_2 x_1 x_2 = 0. \tag{2}$$

We are interested in solutions of this system of equations which lie in the state space. These solutions represent the equilibrium points of our dynamic model.

Step 4 is to solve the model. Factoring out x_1 from the first equation and x_2 from the second, we find four solutions, three at the following coordinates:

$$(0, 0)$$
$$(0, r_2/a_2)$$
$$(r_1/a_1, 0)$$

and the fourth at the intersection of these two lines:

$$a_1 x_1 + b_1 x_2 = r_1$$
$$b_2 x_1 + a_2 x_2 = r_2.$$

See Fig. 4.2 for an illustration. Solving by Cramer's rule yields

$$x_1 = \frac{r_1 a_2 - r_2 b_1}{a_1 a_2 - b_1 b_2}$$

$$x_2 = \frac{a_1 r_2 - b_2 r_1}{a_1 a_2 - b_1 b_2}.$$

If the two lines do not cross inside the state space, then there are only three equilibria. In this case the two species of trees cannot coexist in peaceful equilibrium.

We are interested to know the conditions under which $x_1 > 0$ and $x_2 > 0$. It is reasonable to assume that $a_i > b_i$. The growth rate is

$$r_i x_i - a_i x_i x_i - b_i x_i x_j,$$

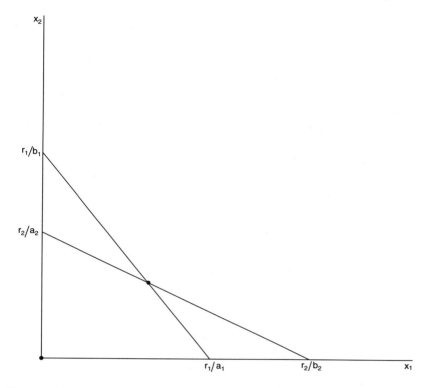

Figure 4.2 Graph of softwoods x_2 versus hardwoods x_1 showing equilibria for the tree problem.

where the first term represents unrestricted growth, the second represents the effect of competition within a population, and the third represents competition between populations. Since the two types of trees do not occupy exactly the same ecological niche, we would suppose that for $x_i = x_j$ the effect of competition within a population would be the stronger. Hence $a_i > b_i$, so

$$a_1 a_2 - b_1 b_2 > 0.$$

The condition for coexistence is, therefore, that

$$r_1 a_2 - r_2 b_1 > 0$$
$$a_1 r_2 - b_2 r_1 > 0,$$

or, in other words,

$$\frac{r_2}{a_2} < \frac{r_1}{b_1} \text{ and } \frac{r_1}{a_1} < \frac{r_2}{b_2},$$

as shown in Fig. 4.2.

Step 5 is to report in plain English the results obtained from our model analysis. It is difficult to do so in this case, because our answer is qualified and the qualifications involve unknown parameters. In order to communicate our results clearly, we would like to find a more tangible interpretation of our conditions for coexistence. Let us reexamine our model formulation to see if we can interpret the meaning of the ratios r_i/a_i and r_i/b_i in some straightforward way.

The parameters r_i measure growth tendency, and the parameters a_i and b_i measure the strength of competition within and between populations, respectively. Thus, the ratios r_i/a_i and r_i/b_i must be measuring the relative strength of growth versus competition. Let us try to go further. In the absence of competition between species, the growth rate is

$$r_i x_i - a_i x_i^2 = x_i (r_i - a_i x).$$

The ratio r_i/a_i represents the equilibrium population level in the absence of competition between species, or the level at which the population will stop growing of its own accord. Similarly, if we neglect the factor of competition within a population, the net growth rate is

$$r_i x_i - b_i x_i x_j = x_i (r_i - b_i x_j).$$

The ratio r_i/b_i thus represents the level of population j necessary to put an end to growth of population i. In light of this, we can now give our analysis results the following concrete interpretation.

For each type of tree (hardwood and softwood) there are two kinds of limits to growth. The first comes from competition with the other type of tree, and the second comes from competition between trees of the same type under crowded conditions. Thus, for each type of tree there is one point where growth will halt itself due to crowding, and another point where the growth of one type of tree will halt the growth of the other type due to competition. The condition for coexistence of both types is that each type reaches the point where it limits its own growth before it reaches the point where it limits the other's growth.

The steady-state analysis of this section leaves one important question unanswered. Given that a dynamic model has an equilibrium solution, will we ever get there? The answer depends on the dynamics of the model. An equilibrium point

$$x_0 = (x_1^0, \ldots, x_n^0)$$

is said to be asymptotically stable (or just stable) if whenever the state variables

$$(x_1(t), \ldots, x_n(t))$$

pass sufficiently close to x_0, they are drawn into the equilibrium. In other words,

$$(x_1(t), \ldots, x_n(t)) \to x_0.$$

Steady-state analysis cannot answer the question of stability, so we will have to defer further discussion of this topic to the next section.

4.2 Dynamical Systems

Dynamical system models are the most commonly used type of dynamic model. In a dynamical system model the forces of change are represented by differential equations. In this section we will focus on the graphical method for obtaining qualitative information about a dynamical system. The emphasis will be on questions of stability.

Example 4.2   The blue whale and fin whale are two similar species that inhabit the same areas. Hence, they are thought to compete. The intrinsic growth rate of each species is estimated at 5% per year for the blue whale and 8% per year for the fin whale. The environmental carrying capacity (the maximum number of whales that the environment can support) is estimated at 150,000 blues and 400,000 fins. The extent to which the whales compete is unknown. In the last 100 years intense harvesting has reduced the whale population to around 5,000 blues and 70,000 fins. Will the blue whale become extinct?

We will use the five-step method. Notice that this problem is very similar to Example 4.1. Step 1 is to ask a question. We will use the number of blue and fin whales as state variables and make the simplest possible assumptions about growth and competition. The question we begin with is this: Can the two populations of whales grow to stable equilibrium starting from their current levels? The results of step 1 are summarized in Figure 4.3.

Step 2 is to select the modeling approach. We will model this problem as a dynamical system.

A *dynamical system* consists of n state variables $(x_1, \ldots, x_n)$ and a system of differential equations

$$\frac{dx_1}{dt} = f_1(x_1, \ldots, x_n)$$

$$\vdots \qquad \vdots \tag{3}$$

$$\frac{dx_n}{dt} = f_n(x_1, \ldots, x_n)$$

Variables: B = number of blue whales
                          F = number of fin whales
                          g_B = growth rate of blue whale population (per year)
                          g_F = growth rate of fin whale population (per year)
                          c_B = effect of competition on blue whales (whales per year)
                          c_F = effect of competition on fin whales (whales per year)

Assumptions: $g_B = .05B(1 - B/150,000)$
                          $g_F = .08F(1 - F/400,000)$
                          $c_B = c_F = \alpha BF$
                          $B \geq 0, F \geq 0$
                          α is a positive real constant

Objective: Determine whether dynamic system can reach stable
                          equilibrium starting from $B = 5,000$, $F = 70,000$

Figure 4.3 Results of step 1 for the whale problem.

defined on the state space $(x_1, \ldots, x_n) \in S$, where S is a subset of $\mathbb{R}^n$. The existence and uniqueness theorem of differential equations states that if $f_1, \ldots, f_n$ have continuous first partial derivatives in a neighborhood of a point

$$x_0 = (x_1^0, \ldots, x_n^0),$$

then there exists a unique solution to this system of differential equations through this initial condition. See any introductory text on differential equations for details (e.g., Hirsch, M. et al., (1974), p. 162).

It is best to think of a solution to a dynamical system as a path through the state space. As long as differentiability assumptions are satisfied, there is a path through each point, and paths cannot cross except at an equilibrium. Let

$$x = (x_1, \ldots, x_n)$$
$$F(x) = (f_1(x), \ldots, f_n(x)).$$

Then the dynamical system equation is

$$\frac{dx}{dt} = F(x). \tag{4}$$

For a path $x(t)$ the derivative dx/dt represents the velocity vector. Hence, for every solution curve $x(t)$ we have that $F(x(t))$ is the velocity vector at each point. The vector field $F(x)$ tells us in what direction and how fast we are moving through the state space. Usually a good idea of the qualitative behavior of a dynamical system in two variables can be obtained by drawing the vector field at selected points. The points where $F(x) = 0$ are the equilibria, and we will pay special attention to the vector field nearby these points.

Step 3 is to formulate the model. Let $x_1 = B$ and $x_2 = F$, and write

$$x_1' = f_1(x_1, x_2)$$
$$x_2' = f_2(x_1, x_2),$$

where

$$f_1(x_1, x_2) = .05x_1(1 - x_1/150, 000) - \alpha x_1 x_2$$
$$f_2(x_1, x_2) = .08x_2(1 - x_2/400, 000) - \alpha x_1 x_2. \tag{5}$$

The state space is

$$S = \{(x_1, x_2) : x_1 \geq 0, x_2 \geq 0\}.$$

Step 4 is to solve the model. We want to sketch a graph of the vector field for this problem. Start out by sketching the level sets $f_1 = 0$ and $f_2 = 0$. The equilibria will be at the intersection of these two. Furthermore, the velocity vectors will be vertical along $f_1 = 0$ ($x_1' = 0$) and horizontal along $f_2 = 0$ ($x_2' = 0$). Draw the velocity vectors along these two curves. Then fill in some of the velocity vectors in between. It helps to remember that (as long as $F(x)$ is continuous, which it usually is) both the length and direction of the vectors change continuously. In fact, for this kind of analysis the length of the velocity vectors is not very important. See Figure 4.4 for the finished graph.

There are four equilibrium solutions, three at

$$(0, 0)$$
$$(150, 000, 0) \tag{6}$$
$$(0, 400, 000)$$

and another at a point whose coordinates depend on α. Our graph assumes that

$$400, 000 < (.05/\alpha).$$

In this case it is easy to see that the equilibrium in the interior is the only stable one. In fact, any solution through a point in the interior of the state space will eventually converge upon this equilibrium. In particular, the solution with initial condition $x_1(0) = 5, 000$ and $x_2(0) = 70, 000$ tends to this equilibrium as $t \to \infty$.

Step 5 is to summarize the results of our model analysis in nonmathematical terms. Based on our analysis, in the absence of further harvesting the whale populations will grow back to their natural levels, and the ecological system will remain in stable equilibrium.

Of course our conclusions are based on some rather broad assumptions. For example, we have assumed that the effect of competition is relatively small. If it were larger (i.e., if $(.05/\alpha) < 400,000$), then the two species could not coexist. It does seem reasonable to make the assumption that α is small, since we know that the two species have coexisted for a long time before we began to harvest them. We have also made several simplifying assumptions about the growth process. The most critical of these is that for very small population levels, the population still tends to increase at the intrinsic growth rate. It is believed that some species have a minimum size (called the *minimum viable population level*) below which the growth rate is negative, insuring the eventual extinction of the

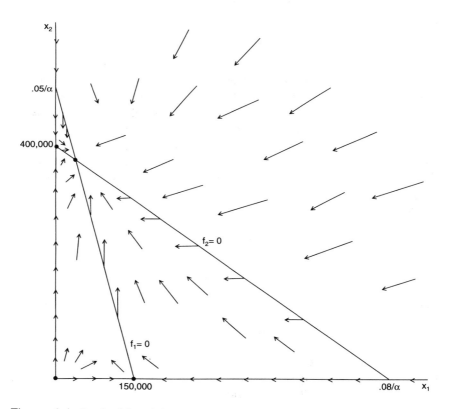

Figure 4.4 Graph of fin whales x_2 versus blue whales x_1 showing vector field for the whale problem.

$$f_1(x_1, x_2) = .05x_1(1 - x_1/150,000) - \alpha x_1 x_2$$
$$f_2(x_1, x_2) = .08x_2(1 - x_2/400,000) - \alpha x_1 x_2$$

species. This assumption would, of course, change the behavior of our dynamical system.

Finally, we address the questions of sensitivity analysis and robustness. First let us consider the sensitivity to the parameter α, for which we have very little information. For any value of $\alpha < 1.25 \times 10^{-7}$ there is a stable equilibrium $x_1 > 0$, $x_2 > 0$ at

$$x_1 = 150,000(8,000,000\alpha - 1)/D$$
$$x_2 = 400,000(1,875,000\alpha - 1)/D,$$

(7)

where

$$D = 15,000,000,000,000\alpha^2 - 1,$$

which we found by Cramer's Rule. For example, if $\alpha = 10^{-7}$, then

$$x_1 = \frac{600,000}{17} \approx 35,294$$
$$x_2 = \frac{6,500,000}{17} \approx 382,353.$$

(8)

The sensitivities at this point are

$$S(x_1, \alpha) = -\frac{21,882,352,927}{6,000,000,000} \approx -3.6$$

and

$$S(x_2, \alpha) = \frac{27}{221} \approx 0.122.$$

The above calculations could be performed either by hand, or by using a computer algebra system. Figure 4.5 illustrates the computation of the sensitivity $S(x_1, \alpha)$ using the computer algebra system MAPLE.

The blue whale population is most sensitive to α. If $\alpha = 10^{-8}$, then

$$x_1 = \frac{276,000,000}{1,997} \approx 138,207$$
$$x_2 = \frac{785,000,000}{1,997} \approx 393,090.$$

Of course we will always have $x_1 < 150,000$ and $x_2 < 400,000$, as is apparent from the graph in Fig. 4.4. But the most important features of this equilibrium are not its coordinates, but rather the fact that an equilibrium exists on $x_1 > 0$, $x_2 > 0$ and is stable. These conclusions remain valid over the entire range of $\alpha < 1.25 \times 10^{-7}$, which we believe to be plausible. Hence, we should say that our main conclusion is not at all sensitive to α. Similarly, our main conclusion is

```
> e1:=(5/100)*(1-x1/150000)-alpha*x2;
                  e1 := 1/20 - 1/3000000 x1 - alpha x2

> e2:=(8/100)*(1-x2/400000)-alpha*x1;
                  e2 := 2/25 - 1/5000000 x2 - alpha x1

> s:=solve({e1=0,e2=0},{x1,x2});
                               - 1 + 1875000 alpha
            s := {x2 = 400000 ---------------------------,
                                                        2
                                - 1 + 15000000000000 alpha


                               - 1 + 8000000 alpha
               x1 = 150000 ---------------------------}
                                                    2
                                - 1 + 15000000000000 alpha

> assign(s);
dx1dalpha:=diff(x1,alpha);
> dx1dalpha:=diff(x1,alpha);
                                  1200000000000
            dx1dalpha := ---------------------------
                                                   2
                          - 1 + 15000000000000 alpha

                               (- 1 + 8000000 alpha) alpha
      - 4500000000000000000 -------------------------------
                                                        2 2
                               (- 1 + 15000000000000 alpha )

> assign(alpha=10^(-7));
> sx1alpha:=dx1dalpha*(alpha/x1);
                                             62
                          sx1alpha := - ----
                                             17

> evalf(sx1alpha);
                          -3.647058824

> stop;
```

Figure 4.5 Calculation of the sensitivity $S(x_1, \alpha)$ for the whale problem using the computer algebra system MAPLE.

not at all sensitive to our data on intrinsic growth rate and carrying capacity, or even to the current whale populations.

Deeper questions of robustness revolve around the assumed form of the functions f_1 and f_2. We assumed that x_1'/x_1 and x_2'/x_2 are linear functions of x_1 and x_2, respectively. These lines represent the point where one species or the other stops

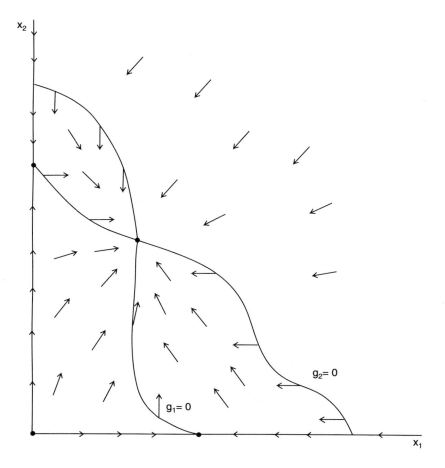

Figure 4.6 Graph of fin whales x_2 versus blue whales x_1 showing vector field for the generalized whale problem.

$$f_1(x_1, x_2) = x_1 g_1(x_1, x_2)$$
$$f_2(x_1, x_2) = x_2 g_2(x_1, x_2)$$

growing. Suppose that we relax this linearity assumption. Let

$$x_1' = x_1 \, g_1(x_1, \, x_2)$$
$$x_2' = x_2 \, g_2(x_1, \, x_2).$$

All of our analysis results would certainly remain true if g_1 and g_2 were not linear, so long as the vector field had the same general features. See Figure 4.6 for an illustration.

4.3   Discrete-Time Dynamical Systems

In some problems it is most natural to model the time variable as being discrete. When this happens the usual differential equations are replaced by their discrete-time analogue: difference equations. The relationship between discrete and continuous dynamics is the relationship between $\Delta x / \Delta t$ and dx/dt, so it is often assumed that the behavior of a dynamical system will be roughly the same whether we assume that time is continuous or discrete. However, this kind of logic overlooks one important point. There is a kind of time delay built into every discrete-time dynamical system, which is the length of the time step Δt. For systems in which the dynamic forces are very strong, this time delay can lead to unexpected results.

Example 4.3   Astronauts in training are required to practice a docking maneuver under manual control. As a part of this maneuver it is required to bring an orbiting spacecraft to rest relative to another orbiting craft. The hand controls provide for variable acceleration and deceleration, and there is a device on board that measures the rate of closing between the two vehicles. The following strategy has been proposed for bringing the craft to rest. First look at the closing velocity. If it is zero, we are done. Otherwise, remember the closing velocity and look at the acceleration control. Move the acceleration control so that it is opposite to the closing velocity (i.e., if closing velocity is positive, we slow down, and we speed up if it is negative) and proportional in magnitude (i.e., we brake twice as hard if we find ourselves closing twice as fast). After a time, look at the closing velocity again and repeat the procedure. Under what circumstances will this strategy be effective?

We will use the five-step method. Let v_n denote the closing velocity observed at time t_n, the time of the nth observation. Let

$$\Delta v_n = v_{n+1} - v_n$$

denote the change in closing velocity as a result of our adjustments. We will

Variables: t_n = time of nth velocity observation (sec)

v_n = velocity at time t_n (m/sec)

c_n = time to make nth control adjustment (sec)

a_n = acceleration after nth adjustment (m/sec^2)

w_n = wait before $(n+1)$th observation (sec)

Assumptions: $t_{n+1} = t_n + c_n + w_n$

$v_{n+1} = v_n + a_{n-1}c_n + a_n w_n$

a_n $= -kv_n$

c_n > 0

w_n ≥ 0

Objective: Determine whether $v_n \to 0$

Figure 4.7 Results of step 1 of the docking problem.

denote the time between observations of the velocity indicator by

$$\Delta t_n = t_{n+1} - t_n.$$

This time interval naturally divides into two parts: the time it takes to adjust the velocity controls and the time between adjustment and the next observation of the velocity indicator. Write

$$\Delta t_n = c_n + w_n,$$

where c_n is the time to adjust the controls and w_n is the waiting time until the next observation. The parameter c_n is a function of astronaut response time, and we are free to choose w_n.

Let a_n denote the acceleration setting after the nth adjustment. Elementary physics yields

$$\Delta v_n = a_{n-1}c_n + a_n w_n.$$

The control law says to set acceleration proportional to $(-v_n)$, hence

$$a_n = -kv_n.$$

The results of step 1 are summarized in Figure 4.7.

Step 2 is to select the modeling approach. We will model this problem as a discrete-time dynamical system.

A discrete-time dynamical system consists of a number of state variables $(x_1, \ldots, x_n)$ defined on the state space $S \subseteq \mathbb{R}^n$ and a system of difference equations

$$\Delta x_1 = f_1(x_1, \ldots, x_n)$$

$$\vdots \qquad \vdots \qquad\qquad (9)$$

$$\Delta x_n = f_n(x_1, \ldots, x_n).$$

Here Δx_n represents the change in x_n over one time step. It is common to take time steps of length 1, which just amounts to selecting appropriate units. If time steps are of variable length, or if the dynamics of the system vary over time, then we include time as a state variable. If we let

$$x = (x_1, \ldots, x_n)$$
$$F = (f_1, \ldots, f_n),$$

then the equations of motion can be written in the form

$$\Delta x = F(x).$$

A solution to this difference equation model is a sequence of points

$$x(0), x(1), x(2), \ldots$$

in the state space with

$$\Delta x(n) = x(n+1) - x(n)$$
$$= F(x(n))$$

for all n. An equilibrium point x_0 is characterized by

$$F(x_0) = 0,$$

and the equilibrium is stable if

$$x(n) \to x_0$$

whenever $x(0)$ is sufficiently close to x_0.

Think of a solution as a sequence of points in the state space. The vector $F(x(n))$ connects the point $x(n)$ to the point $x(n+1)$. A graph of the vector field $F(x)$ can reveal much about the behavior of a discrete-time dynamical system.

Example 4.4   Let $x = (x_1, x_2)$, and consider the difference equation

$$\Delta x = -\lambda x, \qquad\qquad (10)$$

where $\lambda > 0$. What is the behavior of solutions near the equilibrium point $x_0 = (0, 0)$?

Figure 4.8 shows a graph of the vector field $F(x) = -\lambda x$ in the case where $0 < \lambda < 1$. It is clear that $x_0 = (0, 0)$ is a stable equilibrium. Each step moves closer to x_0. Now let us consider what happens when λ becomes larger. Each of the vectors in Fig. 4.8 will stretch as λ increases. For $\lambda > 1$ the vectors are so long that they overshoot the equilibrium. For $\lambda > 2$ they are so long that the terminal point $x(n + 1)$ is actually farther away from $(0, 0)$ than the starting point $x(n)$. In this case x_0 is an unstable equilibrium.

This simple example clearly illustrates the fact that discrete-time dynamical systems do not always behave like their continuous-time analogues. Solutions

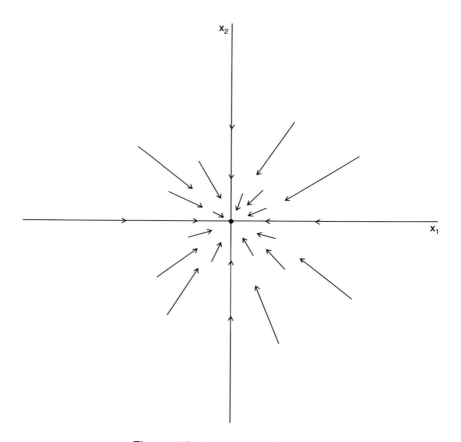

Figure 4.8 Vector field for Example 4.4.

$$f_1(x_1, x_2) = -\lambda x_1$$
$$f_2(x_1, x_2) = -\lambda x_2$$

to the differential equation

$$dx/dt = -\lambda x \qquad (11)$$

are all of the form

$$x(t) = x(0)e^{-\lambda t},$$

and the origin is a stable equilibrium regardless of $\lambda > 0$. The difference in behavior for the analogous difference equation in Eq. (10) is due to the inherent time delay. The approximation

$$dx/dt \approx \Delta x/\Delta t$$

is only valid for small Δt, where the term *small* depends on the sensitivity of x to t. It should only be relied upon in cases where Δx represents a small relative change in x. When this is not the case, the difference in behavior between discrete and continuous systems can be dramatic.

We return now to the docking problem of Example 4.3. Step 3 of the five-step method is to formulate the model. We are modeling the docking problem as a discrete-time dynamical system. From Fig. 4.7 we obtain

$$(v_{n+1} - v_n) = -k\, v_{n-1}\, c_n - k\, v_n w_n.$$

Hence, the change in velocity over the nth time step depends on both v_n and v_{n-1}. To simplify the analysis, let us assume that $c_n = c$ and $w_n = w$ for all n. Then the length of each time step is

$$\Delta t = c + w$$

seconds, and we do not need to include time as a state variable. We do, however, need to include both v_n and v_{n-1}. Let

$$x_1(n) = v_n$$
$$x_2(n) = v_{n-1}.$$

Compute

$$\Delta x_1 = -kwx_1 - kcx_2$$
$$\Delta x_2 = x_1 - x_2. \qquad (12)$$

The state space is $(x_1, x_2) \in \mathbb{R}^2$.

Step 4 is to solve the model. There is one equilibrium point $(0, 0)$ found at the intersection of the two lines

$$kwx_1 + kcx_2 = 0$$
$$x_1 - x_2 = 0.$$

These are the steady-state equations obtained by setting $\Delta x_1 = 0$ and $\Delta x_2 = 0$. Figure 4.9 shows a graph of the vector field

$$F(x) = (-kwx_1 - kcx_2, \; x_1 - x_2).$$

It appears as though solutions will tend toward equilibrium, but it is hard to be sure. If k, c, and w are large, then the equilibrium is probably unstable, but once again it is difficult to tell.

In mathematics we often come across problems we cannot solve. Usually the

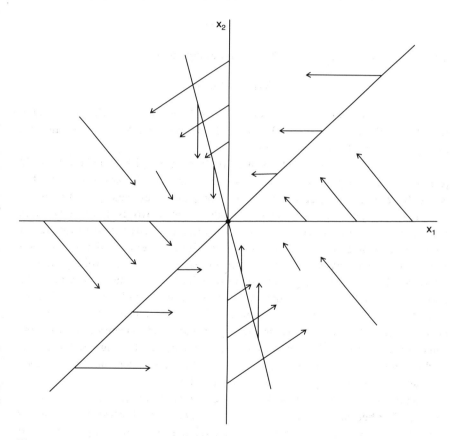

Figure 4.9 Graph of previous velocity x_2 versus current velocity x_1 showing vector field for the docking problem.

$$f_1(x_1, x_2) = -kwx_1 - kcx_2$$
$$f_2(x_1, x_2) = x_1 - x_2$$

best thing to do in such cases is to review our assumptions and consider whether we can reduce the problem to one that we can solve by making a further simplifying assumption. Of course this would be a meaningless and trivial exercise unless the simplified problem had some real significance.

In our docking problem we have expressed the change in velocity Δv_n as the sum of two components. One represents the change in velocity occurring between the time we read the velocity indicator and the time we adjust the acceleration controls. Suppose that we can do this very quickly. In particular, suppose that c is very much smaller than w. If v_n and v_{n-1} are not too different, the approximation

$$\Delta v_n \approx -kwv_n$$

should be reasonably accurate. The difference equation

$$\Delta x_1 = -kwx_1$$

is a familiar one from Example 4.4, and we know that we will get a stable equilibrium for any $kw < 2$. If $kw < 1$, we will approach the equilibrium asymptotically without overshooting.

Step 5 is to answer the analysis question in plain English. Maybe we should just say in plain English that we don't know the answer. However, we probably can do better than that. Let us report that a completely satisfactory solution is not obtainable by elementary graphical methods. In other words, it will take more work using more sophisticated methods to determine exactly the conditions under which the proposed control strategy will work. It does seem that the strategy will be effective in most cases as long as the time interval between control adjustments is not too long and the magnitude of those adjustments is not too large. The problem is complicated by the fact that there is a time delay between reading the velocity indicator and adjusting the controls. Since the actual closing velocity may change during this interval, we are acting on dated and inaccurate information. This adds an element of uncertainty to our calculations. If we ignore the effects of this time delay (which may be permissible if the delay is small), we can then draw some general conclusions, which are as follows.

The control strategy will work so long as the control adjustments are not too violent. Furthermore, the longer the interval between adjustments, the lighter those adjustments must be. In addition, the relationship is one of proportion. If we go twice as long between adjustments, we can only use half as much control. To be specific, if we adjust the controls once every 10 seconds, then we can only set the acceleration controls at 1/10 of the velocity setting to avoid overshooting the target velocity of zero. In order to allow for human and equipment error, we should actually set the controls somewhat lower, say 1/15 or 1/20 of velocity. More frequent adjustments require more frequent observations of the closing velocity indicator and more concentration on the part of the operator, but they do allow for the successful administration of more thrusting power under control. Presumably this would be advantageous.

Normally we would conclude our discussion of this problem with a fairly comprehensive sensitivity analysis. In view of the fact that we have not yet found a way to solve this problem, we will defer that discussion to a later chapter.

4.4 Exercises

1. Reconsider the tree problem of Example 4.1. Assume that

$$\frac{r_2}{a_2} < \frac{r_1}{b_1} \text{ and } \frac{r_1}{a_1} < \frac{r_2}{b_2}$$

so that the situation is as pictured in Fig. 4.2.
(a) Draw the vector field for this model.
(b) Classify each of the four equilibrium points as stable or unstable.
(c) Can the two species of trees coexist in stable equilibrium?
(d) Suppose that a logging operation removes all but a few of the valuable hardwood trees in this stand of forest. What does this model predict about the future of the two species of trees?

2. Reconsider the tree problem of Example 4.1, but now assume that

$$\frac{r_2}{a_2} < \frac{r_1}{b_1} \text{ and } \frac{r_1}{a_1} \geq \frac{r_2}{b_2}.$$

(a) Locate each of the equilibrium points (x_1, x_2) in the state space $x_1 \geq 0$, $x_2 \geq 0$.
(b) Draw the vector field for this case.
(c) Classify each equilibrium as stable or unstable.
(d) Suppose that we start out with an equal amount of hardwood and softwood trees. What does this model predict about the future of the two species?

3. Repeat Exercise 2, but now assume that

$$\frac{r_2}{a_2} \geq \frac{r_1}{b_1} \text{ and } \frac{r_1}{a_1} < \frac{r_2}{b_2}.$$

4. In the whale problem of Example 4.2 we used a logistic model of population growth, where the growth rate of population P in the absence of interspecies competition is

$$g(P) = rP - aP^2.$$

In this problem we will be using the simpler growth model

$$g(P) = rP.$$

(a) Can both species of whales coexist? Use the five-step method, and model as a dynamical system in steady-state.
(b) Draw the vector field for this model. Indicate the location of each equilibrium point.
(c) Classify each equilibrium point in the state space as stable or unstable.
(d) Suppose that there are currently 5,000 blue whales and 70,000 fin whales. What does this model predict about the future of the two species?

5. In the whale problem of Example 4.2 we used a logistic model of population growth, where the growth rate of population P in the absence of interspecies competition is

$$g(P) = P(r - aP).$$

In this problem we will be using a more complex model,

$$g(P) = (P - c)(r - aP),$$

in which the parameter c represents a minimum viable population level below which the growth rate is negative. Assume that $\alpha = 10^{-7}$ and that the minimum viable population level is 3,000 for blue whales and 15,000 for fin whales.

(a) Can the two species of whales coexist? Use the five-step method, and model as a dynamical system in steady-state.
(b) Sketch the vector field for this model. Classify each equilibrium point as stable or unstable.
(c) Assuming that there are currently 5,000 blue whales and 70,000 fin whales, what does this model predict about the future of the two populations?
(d) Suppose that we have underestimated the minimum viable population for the blue whale, and that it is actually closer to 10,000. Now what happens to the two species?

6. Reconsider the whale problem of Example 4.2, and assume that $\alpha = 10^{-8}$. In this problem we will investigate the effects of harvesting on the two whale populations. Assume that a level of effort E boat-days will result in the annual harvest of qEx_1 blue whales and qEx_2 fin whales, where the parameter q (catchability) is assumed to equal approximately 10^{-5}.

(a) Under what conditions can both species continue to coexist in the presence of harvesting? Use the five-step method, and model as a dynamical system in steady-state.

(b) Draw the vector field for this problem, assuming that the conditions identified in part (a) are satisfied.

(c) Find the minimum level of effort required to reduce the fin whale population to its current level of around 70,000 whales. Assume that we started out with 150,000 blue whales and 400,000 fin whales before mankind began to harvest them.

(d) Describe what would happen to the two populations if harvesting were allowed to continue at the level of effort identified in part (c). Draw the vector field in this case. This is the situation which led the IWC to call for an international ban on whaling.

7. One of the favorite foods of the blue whale is called krill. These tiny shrimp-like creatures are devoured in massive amounts to provide the principal food source for the huge whales. The maximum sustainable population for krill is 500 tons/acre. In the absence of predators, in uncrowded conditions, the krill population grows at a rate of 25% per year. The presence of 500 tons/acre of krill increases blue whale population growth rate by 2% per year, and the presence of 150,000 blue whales decreases krill growth rate by 10% per year.

(a) Determine whether the whales and the krill can coexist in equilibrium. Use the five-step method, and model as a dynamical system in steady-state.

(b) Draw the vector field for this problem. Classify each equilibrium point in the state space as stable or unstable.

(c) Describe what happens to the two populations over time. Assume that we start off with 5,000 blue whales and 750 tons/acre of krill.

(d) How sensitive are your conclusions in part (c) to the assumption of a 25% growth rate per year for krill?

8. Two armies are to engage in battle. The red army enjoys a three-to-one numerical superiority, but the blue army is better trained and better equipped. Let R and B denote the force levels of red and blue forces. The Lanchester model of combat states that

$$R' = -aB - bRB$$
$$B' = -cR - dRB,$$

where the first term accounts for direct fire (aimed at a specific target) and the second term accounts for attrition due to area fire (e.g., artillery). We are assuming that weapon effectiveness is higher for blue than for red, i.e., $a > c$ and $b > d$. But what kind of edge in weapon effectiveness would be necessary to counteract a 3:1 numerical superiority?

(a) Assuming that $a = \lambda c$ and $b = \lambda d$ for some $\lambda > 1$, determine the approximate lower bound on λ necessary for blue to win the war. Use the five-step method, and model as a dynamical system.

(b) In part (a) you assumed red had an $n{:}1$ numerical superiority. Discuss the sensitivity of your results in part (a) to the parameter $n \in (2, 5)$.

9. The following simple model is intended to represent the dynamics of supply and demand. Let P denote the selling price of a certain product and Q the quantity of this product being produced. The supply curve $Q = f(P)$ tells how much should be produced at a given price to maximize profit. The demand curve $P = g(Q)$ tells what price buyers should pay given a certain level of production, in order to maximize their utility.

(a) Select a specific product, and make an educated guess as to the form of the supply curve $Q = f(P)$ and the demand curve $P = g(Q)$.

(b) Use the results of part (a) to determine the equilibrium levels of P and Q.

(c) Formulate a dynamic model based on the assumption that P will tend toward the level dictated by the demand curve, while Q will tend to the level given by the supply curve.

(d) According to your model, is the (P, Q) equilibrium stable? Does it matter whether you assume a discrete-time or continuous-time model? (Economists usually assume a discrete-time model in order to represent the effect of a time delay.)

(e) Perform a sensitivity analysis for the assumptions you made in part (a). Consider the question of stability.

10. A population of 100,000 members is subject to a disease that is seldom fatal and leaves the victim immune to future infections by this disease. Infection can only occur when a susceptible person comes in direct contact with an infectious person. The infectious period lasts approximately three weeks. Last week there were 18 new cases of the disease reported. This week there were 40 new cases. It is estimated that 30% of the population is immune due to previous exposure.

(a) What is the eventual number of people who will become infected? Use the five-step method, and model as a discrete-time dynamical system.

(b) Estimate the maximum number of new cases in any one week.

(c) Conduct a sensitivity analysis to investigate the effect of any assumptions you made in part (a) which were not supported by hard data.

(d) Perform a sensitivity analysis for the number (18) of cases reported last week. It is thought by some that in early weeks the epidemic might be underreported.

11. Reconsider the docking problem of Example 4.3, and now assume that $c = 5$ sec, $w = 10$ sec, and $k = 0.02$.

(a) Assuming an initial closing velocity of 50 m/sec, calculate the sequence of velocity observations $v_0, v_1, v_1, \ldots$, predicted by the model. Is the docking procedure successful?

(b) An easier way to compute the solution in part (a) is to use the iteration function $G(x) = x + F(x)$, with the property that $x(n + 1) = G(x(n))$. Compute the iteration function for this problem, and use it to repeat the calculation in part (a).

(c) Calculate the solution $x(1)$, $x(2)$, $x(3)$, ..., starting at $x(0) = (1, 0)$. Repeat, starting at $x(0) = (0, 1)$. What happens as $n \to \infty$? What does this imply about the stability of the equilibrium $(0, 0)$? [Hint: Every possible initial condition $x(0) = (a, b)$ can be written as a linear combination of the vectors $(1, 0)$ and $(0, 1)$, and $G(x)$ is a linear function of x.]

(d) Are there any states x for which $G(x) = \lambda x$ for some real λ? If so, what happens to the system if we start with this initial condition?

Further Reading

Bailey, N. (1975). *The Mathematical Theory of Infectious Disease*, Hafner Press, New York.

Casstevens, T. *Population Dynamics of Governmental Bureaus*, UMAP module 494.

Clark, C. (1976). *Mathematical Bioeconomics: The Optimal Management of Renewable Resources*, Wiley, New York.

Giordano, F. and Leja, S. *Competitive Hunter Models*, UMAP module 628.

Greenwell, R. *Whales and Krill: A Mathematical Model*, UMAP module 610.

Greenwell, R. and Ng, H. *The Ricker Salmon Model*, UMAP module 653.

Hirsch, M. and Smale, S. (1974). *Differential Equations, Dynamical Systems, and Linear Algebra*, Academic Press, New York.

Horelick, B., Koont, S. and Gottleib, S. *Population Growth and the Logistic Curve*, UMAP module 68.

Horelick, B. and Koont, S. *Epidemics*, UMAP module 73.

Lanchester, F. (1956). Mathematics in Warfare, *The World of Mathematics*, Vol. 4, pp. 1240–1253.

May, R. (1976). *Theoretical Ecology: Principles and Applications*, Saunders, Philadelphia.

Morrow, J. *The Lotka–Volterra Predator–Prey Model*, UMAP module 675.

Sherbert, D. *Difference Equations with Applications*, UMAP module 322.

Tuchinsky, P. *Management of a Buffalo Herd*, UMAP module 207.

Chapter Five

Analysis of Dynamic Models

In this chapter we consider some of the most broadly applicable techniques for the analysis of discrete-time and continuous-time dynamical systems. Except for a few special cases, these methods do not yield exact analytical solutions. Such exact analytical methods are more properly treated in a course on differential equations. In any case, most dynamical system models that arise in practice are not amenable to exact solution by any known technique. In this chapter we will present techniques that can be applied to the analysis of almost any dynamical system model. These methods can provide important qualitative information about the behavior of dynamical systems, even when exact analytic solutions are not obtainable.

5.1 Eigenvalue Methods

When the equations of a dynamic model are linear, it is possible to obtain an exact analytical solution. While linear dynamics are rare in real life, the majority of dynamic models can be approximated by linear models at least locally. Such linear approximations, especially in the neighborhood of an isolated equilibrium point, provide the basis for many of the most important analytical techniques available for dynamic modeling.

Example 5.1   Reconsider the tree problem of Example 4.1. Assume that hardwoods grow at a rate of 10% per year and softwoods at a rate of 25% per year. An acre of forest land can support about 10,000 tons of hardwoods, or 6,000 tons of softwoods. The extent of competition has not been numerically determined. Can both types of trees coexist in stable equilibrium?

Step 1 of the five-step method was laid out in Fig. 4.1. In this particular case we have

$$r_1 = 0.10$$
$$r_2 = 0.25$$
$$a_1 = 0.10/10,000$$
$$a_2 = 0.25/6,000.$$

Step 2 is to select the modeling approach, including a method of analysis. We will analyze this nonlinear dynamical system by the eigenvalue method.

> Suppose we are given a dynamical system $x' = F(x)$ where $x = (x_1, \ldots, x_n)$ is an element of the state space $S \subseteq \mathbb{R}^n$ and $F = (f_1, \ldots, f_n)$. A point $x_0 \in S$ is an equilibrium or steady state if and only if $F(x_0) = 0$. There is a theorem that states that an equilibrium point x_0 is asymptotically stable if the matrix
>
> $$A = \begin{pmatrix} \partial f_1/\partial x_1 (x_0) & \cdots & \partial f_1/\partial x_n (x_0) \\ \vdots & & \vdots \\ \partial f_n/\partial x_1 (x_0) & \cdots & \partial f_n/\partial x_n (x_0) \end{pmatrix} \qquad (1)$$
>
> has eigenvalues with all negative real parts. If any eigenvalue has a positive real part, then the equilibrium is unstable. In the remaining cases (pure imaginary eigenvalues) the test is inconclusive (Hirsch, M. et al., (1974), p. 187).
>
> The eigenvalue method is based on a linear approximation. Even if $x' = F(x)$ is not linear, we will have
>
> $$F(x) \approx A(x - x_0)$$
>
> in the neighborhood of the equilibrium point. This is the same sort of linear approximation you saw in one-variable calculus, except that now the derivative of F is represented by a matrix. Some authors will call this matrix DF in analogy to the one-variable derivative. The linear approximation is good enough so that if the origin is a stable equilibrium of $x' = Ax$ (i.e., the point x_0 is a stable equilibrium of $x' = A(x - x_0)$), then x_0 is a stable equilibrium of $x' = F(x)$ as well. Therefore, it is enough to understand the eigenvalue test in the case of a linear system.
>
> Undoubtedly you solved some linear systems of differential equations in your introductory differential equations course, and you probably learned about the relation between solutions and eigenvalues. For example, if $Au = \lambda u$

(i.e., u is an eigenvector of A belonging to the eigenvalue λ), then $x(t) = ue^{\lambda t}$ is a solution to the initial value problem

$$x' = Ax, \quad x(0) = u.$$

It is actually possible to write down the general solution to an $n \times n$ system of linear differential equations, although it is rather messy and requires a lot of linear algebra. One good thing that comes from doing so, however, is a general description of solution behavior. This theorem says that for any solution $x(t)$ to the differential equation $x' = Ax$, each coordinate is a linear combination of terms that look like one of

$$t^k e^{at} \cos(bt), \quad t^k e^{at} \sin(bt),$$

where $a \pm ib$ is an eigenvalue of A (if the eigenvalue is real, then $b = 0$) and k is a nonnegative integer less than n. From this general description it is easy to calculate that the origin is an asymptotically stable equilibrium of the system $x' = Ax$ if and only if every eigenvalue $a \pm ib$ has $a < 0$ (Hirsch, M. et al., (1974), p. 135).

Of course, a successful application of the eigenvalue method requires us to be able to compute the eigenvalues. In simple cases (e.g., on $\mathbb{R}^2$) it will be possible to compute eigenvalues by hand, or possibly with the aid of a computer algebra system. Otherwise we will have to rely on approximate methods. Fortunately, numerical analysis software packages do exist to compute the eigenvalues of an $n \times n$ matrix, and these are effective in most cases (e.g., see Press, et al., (1986)).

Returning to Example 5.1, recall from Section 4.1 that there is an equilibrium at the point

$$x_1 = (r_1 a_2 - r_2 b_1)/D$$
$$x_2 = (a_1 r_2 - b_2 r_1)/D,$$

where

$$D = (a_1 a_2 - b_1 b_2).$$

We have now specified values for a_1, a_2, r_1, and r_2, but not for b_1 and b_2. We will, however, continue to assume that $b_i < a_i$. For the moment let us take $b_i = a_i/2$. Then the coordinates of the equilibrium point are $x_0 = (x_1^0, x_2^0)$, where

$$x_1^0 = \frac{28,000}{3} \approx 9,333$$

$$x_2^0 = \frac{4,000}{3} \approx 1,333.$$

(2)

The dynamical system equations are $x' = F(x)$, where $F = (f_1, f_2)$ and

$$
\begin{aligned}
f_1(x_1, x_2) &= 0.10x_1 - \frac{0.10}{10,000}x_1^2 - \frac{0.05}{10,000}x_1x_2 \\
f_2(x_1, x_2) &= 0.25x_2 - \frac{0.25}{6,000}x_2^2 - \frac{0.125}{6,000}x_1x_2.
\end{aligned}
\tag{3}
$$

The partial derivatives are

$$
\begin{aligned}
\frac{\partial f_1}{\partial x_1} &= \frac{20,000 - x_2}{200,000} - \frac{x_1}{50,000} \\
\frac{\partial f_1}{\partial x_2} &= \frac{-x_1}{200,000} \\
\frac{\partial f_2}{\partial x_1} &= \frac{-x_2}{48,000} \\
\frac{\partial f_2}{\partial x_2} &= \frac{-x_1}{48,000} - \frac{x_2}{12,000} + \frac{1}{4}.
\end{aligned}
\tag{4}
$$

Evaluating the partial derivatives in Eq. (4) at the equilibrium point in Eq. (2) and substituting back into Eq. (1), we obtain

$$
A = \begin{pmatrix} -7/75 & -7/150 \\ -1/36 & -1/18 \end{pmatrix}.
\tag{5}
$$

The eigenvalues of this 2×2 matrix can be computed as the roots of the equation

$$
\begin{vmatrix} \lambda + 7/75 & 7/150 \\ 1/36 & \lambda + 1/18 \end{vmatrix} = 0.
$$

Evaluating the determinant, we obtain the equation

$$
\frac{1800\lambda^2 + 268\lambda + 7}{1800} = 0,
$$

and then we obtain

$$
\lambda = \frac{-67 \pm \sqrt{1339}}{900}.
$$

Since both eigenvalues have negative real part, the equilibrium is stable.

The eigenvalue test for continuous-time dynamical systems involves quite a bit of computation. This is an appropriate application for a computer algebra system. Figure 5.1 illustrates the use of the computer algebra system MATHEMATICA to perform the computations for step 4 of the present problem.

Finally, step 5. We have found that hardwoods and softwoods can coexist in stable equilibrium. There will be approximately 9,300 tons/acre of hardwoods

```
In[1]:= f1=x1/10-(x1^2/10)/10000-(5 x1 x2/100)/10000
```

$$\text{Out[1]}= \frac{x1}{10} - \frac{x1^2}{100000} - \frac{x1\ x2}{200000}$$

```
In[2]:= f2=25 x2/100-(25 x2^2/100)/6000-(125 x1 x2/1000)/6000
```

$$\text{Out[2]}= \frac{x2}{4} - \frac{x1\ x2}{48000} - \frac{x2^2}{24000}$$

```
In[3]:= s=Solve[{f1/x1==0,f2/x2==0},{x1,x2}]
```

$$\text{Out[3]}= \{\{x1 \to \frac{28000}{3}, \ x2 \to \frac{4000}{3}\}\}$$

```
In[4]:= f={f1,f2}
```

$$\text{Out[4]}= \{\frac{x1}{10} - \frac{x1^2}{100000} - \frac{x1\ x2}{200000}, \ \frac{x2}{4} - \frac{x1\ x2}{48000} - \frac{x2^2}{24000}\}$$

```
In[5]:= df={D[f,x1],D[f,x2]}
```

$$\text{Out[5]}= \{\{\frac{1}{10} - \frac{x1}{50000} - \frac{x2}{200000}, \ \frac{-x2}{48000}\}, \ \{\frac{-x1}{200000}, \ \frac{1}{4} - \frac{x1}{48000} - \frac{x2}{12000}\}\}$$

```
In[6]:= df/.s
```

$$\text{Out[6]}= \{\{\{-(\frac{7}{75}), \ -(\frac{1}{36})\}, \ \{-(\frac{7}{150}), \ -(\frac{1}{18})\}\}\}$$

```
In[7]:= Eigenvalues[{{-7/75,-1/36},{-7/150,-1/18}}]
```

$$\text{Out[7]}= \{\frac{-268 + 4\ \text{Sqrt}[1339]}{3600}, \ \frac{-268 - 4\ \text{Sqrt}[1339]}{3600}\}$$

Figure 5.1 Calculations for step 4 of the tree problem using the computer algebra system MATHEMATICA.

and 1,300 tons/acre of softwoods in a mature, stable forest. These conclusions are based on certain plausible assumptions about the degree of competition between the two types of trees. A sensitivity analysis will be conducted to determine the effect of these assumptions on our broad conclusions.

For the sensitivity analysis, we will still assume that $b_i = ta_i$, but we will relax the assumption that $t = 1/2$. The conditions

$$b_i < a_i$$

$$(r_i/a_i) < (r_j/b_j)$$

imply that $0 < t < 0.6$. The coordinates of the equilibrium point (x_1^0, x_2^0) are

$$
\begin{aligned}
x_1^0 &= (10,000 - 6,000t)/(1 - t^2) \\
x_2^0 &= (6,000 - 10,000t)/(1 - t^2).
\end{aligned}
\tag{6}
$$

The differential equations of this system are $x_i' = f_i(x_1, x_2)$, where

$$
\begin{aligned}
f_1(x_1, x_2) &= 0.10x_1 - \frac{0.10x_1^2}{10,000} - \frac{0.10tx_1x_2}{10,000} \\
f_2(x_1, x_2) &= 0.25x_2 - \frac{0.25x_2^2}{6,000} - \frac{0.25tx_1x_2}{6,000},
\end{aligned}
\tag{7}
$$

and the partial derivatives are

$$
\begin{aligned}
\frac{\partial f_1}{\partial x_1} &= \frac{10,000 - tx_2}{100,000} - \frac{x_1}{50,000} \\
\frac{\partial f_1}{\partial x_2} &= \frac{-tx_1}{100,000} \\
\frac{\partial f_2}{\partial x_1} &= \frac{-tx_2}{24,000} \\
\frac{\partial f_2}{\partial x_2} &= \frac{-tx_1}{24,000} - \frac{x_2}{12,000} + \frac{1}{4}.
\end{aligned}
\tag{8}
$$

Evaluating the partial derivatives in Eq. (8) at the equilibrium point in Eq. (6) and substituting back into Eq. (1) yields

$$
A = \begin{pmatrix}
\dfrac{5 - 3t}{50(t^2 - 1)} & \dfrac{t(5 - 3t)}{50(t^2 - 1)} \\[3mm]
\dfrac{t(3 - 5t)}{12(t^2 - 1)} & \dfrac{3 - 5t}{12(t^2 - 1)}
\end{pmatrix}.
\tag{9}
$$

The characteristic equation we must solve to find the eigenvalues is

$$
\left[\lambda - \frac{5 - 3t}{50(t^2 - 1)}\right]\left[\lambda - \frac{3 - 5t}{12(t^2 - 1)}\right] - \left[\frac{t(3 - 5t)}{12(t^2 - 1)}\right]\left[\frac{t(5 - 3t)}{50(t^2 - 1)}\right] = 0.
\tag{10}
$$

Solving Eq. (10) for λ yields two roots:

$$\lambda_1 = \frac{143t - 105 + \sqrt{9,000t^4 - 20,400t^3 + 20,449t^2 - 9,630t + 2,025}}{600(1 - t^2)}$$

$$\lambda_2 = \frac{143t - 105 - \sqrt{9,000t^4 - 20,400t^3 + 20,449t^2 - 9,630t + 2,025}}{600(1 - t^2)}.$$

(11)

Figure 5.2 illustrates the use of the computer algebra system MAPLE to compute the eigenvalues for this problem. Computer algebra systems are especially useful for problems like this one, where the calculations become complicated and there is an increased risk of error when doing all of the algebra by hand. Most computer algebra systems also include a graphing utility. The combination of graphics and algebra is important in problems such as the present one. Drawing a graph is often the easiest way to solve an inequality.

Figure 5.3 shows a graph of λ_1 and λ_2 versus t over the interval $0 < t < 0.6$. From this graph we can see that λ_1, λ_2 are always negative, so that the equilibrium is stable regardless of the strength of competition. (If you did Exercise 1 of Chapter 4, you probably drew the same conclusion from a graphical analysis.)

5.2   Eigenvalue Methods for Discrete Systems

The methods of the previous section apply only to continuous-time dynamical systems. In this section we will present analogous methods for the stability analysis of discrete-time dynamical systems. Once again the basis for our analysis is a linear approximation, together with a calculation of eigenvalues.

Example 5.2   Reconsider the docking problem of Example 4.3, and assume now that it takes 5 seconds to make the control adjustments, and another 10 seconds until we can return from other tasks to observe the velocity indicator once again. Under these conditions, will our strategy for matching velocities be successful?

Step 1 of the five-step method was summarized in Fig. 4.7. Now we will assume $c_n = 5$ and $w_n = 10$. For the moment we will set $k = 0.02$, and then later we will perform a sensitivity analysis on k.

Step 2 is to select the modeling approach, including the method of solution. We will use an eigenvalue method.

Given a discrete-time dynamical system

$$\Delta x = F(x),$$

```
> with(linalg);
> f1:=x1/10-(x1^2/10)/10000-(t*x1*x2/10)/10000;
                           2
          f1 := 1/10 x1 - 1/100000 x1  - 1/100000 t x1 x2

> f2:=25*x2/100-(25*x2^2/100)/6000-(25*t*x1*x2/100)/6000;
                           2
          f2 := 1/4 x2 - 1/24000 x2  - 1/24000 t x1 x2

> df1dx1:=diff(f1,x1);
          df1dx1 := 1/10 - 1/50000 x1 - 1/100000 t x2

> df1dx2:=diff(f1,x2);
                df1dx2 := - 1/100000 t x1
> df2dx1:=diff(f2,x1);
                df2dx1 := - 1/24000 t x2
> df2dx2:=diff(f2,x2);
          df2dx2 := 1/4 - 1/12000 x2 - 1/24000 t x1

> s:=solve({f1/x1=0,f2/x2=0},{x1,x2});
                    - 5 + 3 t              - 3 + 5 t
          s := {x1 = 2000 ---------, x2 = 2000 ---------}
                               2                    2
                         - 1 + t              - 1 + t
> assign(s);
> A:=array([[df1dx1,df1dx2],[df2dx1,df2dx2]]);
   A :=             - 5 + 3 t         t (- 3 + 5 t)            t (- 5 + 3 t)
          [1/10 - 1/25 --------- - 1/50 -------------, - 1/50 -------------]
                            2                 2                      2
                      - 1 + t           - 1 + t                - 1 + t

               t (- 3 + 5 t)           - 3 + 5 t         t (- 5 + 3 t)
          [- 1/12 -------------, 1/4 - 1/6 --------- - 1/12 -------------]
                       2                        2                 2
                 - 1 + t                  - 1 + t            - 1 + t
> eigenvals(A);
                       2                          4            3 1/2
       - 286 t + 210 + 2 (20449 t  - 9630 t + 2025 + 9000 t  - 20400 t )
   1/2 -----------------------------------------------------------------,
                                     2
                               600 t  - 600

                       2                          4            3 1/2
       - 286 t + 210 - 2 (20449 t  - 9630 t + 2025 + 9000 t  - 20400 t )
   1/2 -----------------------------------------------------------------
                                     2
                               600 t  - 600
```

Figure 5.2 Calculations for sensitivity analysis in the tree problem using the computer algebra system MAPLE.

where $x = (x_1, \ldots, x_n)$ and $F = (f_1, \ldots, f_n)$, let us define the iteration function

$$G(x) = x + F(x).$$

The sequence $x(0)$, $x(1)$, $x(2)$, ..., is a solution to this system of difference equations if and only if

$$x(n + 1) = G(x(n))$$

for all n. An equilibrium point x_0 is characterized by the fact that x_0 is a fixed point of the function $G(x)$, i.e., $G(x_0) = x_0$.

There is a theorem that states that an equilibrium point x_0 is (asymptotically) stable if every eigenvalue of the matrix of partial derivatives

$$A = \begin{pmatrix} \partial g_1/\partial x_1\,(x_0) & \cdots & \partial g_1/\partial x_n\,(x_0) \\ \vdots & & \vdots \\ \partial g_n/\partial x_1\,(x_0) & \cdots & \partial g_n/\partial x_n\,(x_0) \end{pmatrix} \tag{12}$$

has absolute value less than one. If the eigenvalue is complex $a \pm ib$, then by absolute value we mean the complex absolute value $\sqrt{a^2 + b^2}$. This

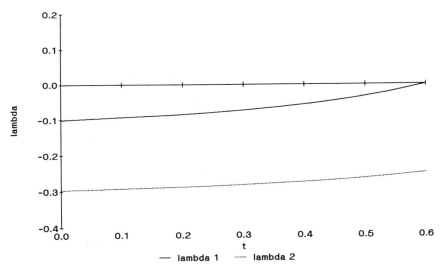

Figure 5.3 Graph of the eigenvalues λ_1 and λ_2 versus the parameter t in the tree problem.

$$\lambda_1 = \frac{143t - 105 + \sqrt{9,000t^4 - 20,400t^3 + 20,449t^2 - 9,630t + 2,025}}{600(1 - t^2)}$$

$$\lambda_2 = \frac{143t - 105 - \sqrt{9,000t^4 - 20,400t^3 + 20,449t^2 - 9,630t + 2,025}}{600(1 - t^2)}$$

simple test for stability is analogous to the eigenvalue test for continuous time dynamical systems presented in the preceding section (Hirsch, M. et al., (1974), p. 280).

As in the continuous case, the eigenvalue method for discrete-time dynamical systems is based on a linear approximation. Even if the iteration function $G(x)$ is not linear, we will have

$$G(x) \approx A(x - x_0)$$

in the neighborhood of the equilibrium point x_0. In other words, the behavior of the iteration function G in the neighborhood of the equilibrium point x_0 is approximately the same as the behavior of the linear function Ax near the origin. Therefore, the behavior of our original nonlinear system in the neighborhood of x_0 is approximately the same as the behavior of the linear discrete-time dynamical system defined by the iteration function

$$x(n + 1) = Ax(n)$$

in the neighborhood of the origin. The linear approximation is good enough so that if the origin is a stable equilibrium of the linear system, then x_0 is a stable equilibrium of the original nonlinear system. So it only remains to discuss the conditions for stability of the linear system.

A matrix A is called a *linear contraction* if $A^n x \rightarrow 0$ for every x. There is a theorem that states that if every eigenvalue of a matrix A has absolute value less than one, then A is a linear contraction (Hirsch, M. et al., (1974), p. 279). It follows that the origin is a stable equilibrium of the discrete-time dynamical system with iteration function A whenever the eigenvalues of A all have absolute value less than one. We illustrate the proof of this result in a simple case. Suppose that $Ax = \lambda x$ for *all* x. Then λ is an eigenvalue of A, and every nonzero vector x is an eigenvector belonging to λ. In this simple case we will always have

$$x(n + 1) = Ax(n) = \lambda x(n),$$

so that the origin is a stable equilibrium if and only if $|\lambda| < 1$.

Now we return to the docking problem. Step 3 is to formulate the model as necessary for the application of the techniques identified in step 2. In this case we already have a linear system with equilibrium $x_0 = (0, 0)$. The iteration function is

$$G(x_1, x_2) = x + F(x_1, x_2) = (g_1, g_2)$$

where

$$g_1(x_1, x_2) = .8x_1 - .1x_2$$
$$g_2(x_1, x_2) = x_1.$$

Moving on to step 4, we calculate

$$\begin{vmatrix} \lambda - .8 & .1 \\ -1 & \lambda - 0 \end{vmatrix} = 0,$$

or $\lambda^2 - .8\lambda + .1 = 0$, from which we obtain

$$\lambda = \frac{4 \pm \sqrt{6}}{10}.$$

There are $n = 2$ distinct eigenvalues, and both are real and lie between -1 and $+1$. Hence the equilibrium $x_0 = 0$ is stable, so we will get $x(t) \to (0,\ 0)$ for any initial condition.

Step 5 is to state our results in plain English. We assumed 15 seconds between control adjustments: 5 seconds to make the adjustment and 10 seconds slack time. Using a correction factor of 1:50, we can guarantee success for our proportional method of control. In practical terms, a correction factor of 1:50 means that if the velocity indicator reads 50 m/sec we will set the acceleration controls for -1 m/sec^2. If the reading is 25 m/sec, we set the controls at -0.5 m/sec^2, and so on.

What follows is a sensitivity analysis for the parameter k. For a general k the iteration function is given by $G = (g_1,\ g_2)$, where

$$g_1(x_1,\ x_2) = (1 - 10k)x_1 - 5kx_2$$
$$g_2(x_1,\ x_2) = x_1,$$

which leads to the characteristic equation

$$\lambda^2 - (1 - 10k)\lambda + 5k = 0.$$

The eigenvalues are

$$\lambda_1 = \frac{(1 - 10k) + \sqrt{(1 - 10k)^2 - 20k}}{2}$$

$$\lambda_2 = \frac{(1 - 10k) - \sqrt{(1 - 10k)^2 - 20k}}{2}.$$

(13)

The quantity under the radical in Eq. (13) becomes negative between

$$k_1 = \frac{4 - \sqrt{12}}{20} \approx 0.027$$

and

$$k_2 = \frac{4 + \sqrt{12}}{20} \approx 0.373.$$

Figure 5.4 shows a graph of λ_1 and λ_2 over the interval $0 < k \leq k_1$. From the graph we can see that both eigenvalues have absolute value less than one, so the equilibrium $(0, 0)$ is stable over this entire range of k. For $k_1 < k < k_2$, both eigenvalues are complex, and the condition for stability is that

$$\left[\frac{(1 - 10k)^2}{2} \right] + \left[\frac{\sqrt{20k - (1 - 10k)^2}}{2} \right] < 1, \tag{14}$$

which reduces to $k < 1/5$. Figure 5.5 shows a graph of λ_1 and λ_2 for $k \geq k_2$. It is easy to see that the smaller eigenvalue, λ_2, has absolute value greater than one for all such k. To summarize, the method will achieve matched velocities as long as $k < 0.2$, or at least a 1:5 correction factor. Of course, it is of some interest to know which value of k is the most efficient. We will leave this problem for the exercises.

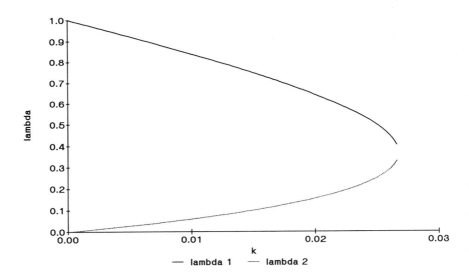

Figure 5.4 Graph of the eigenvalues λ_1 and λ_2 versus control parameter k in the docking problem: case $0 < k \leq k_1$.

$$\lambda_1 = \frac{(1 - 10k) + \sqrt{(1 - 10k)^2 - 20k}}{2}$$

$$\lambda_2 = \frac{(1 - 10k) - \sqrt{(1 - 10k)^2 - 20k}}{2}$$

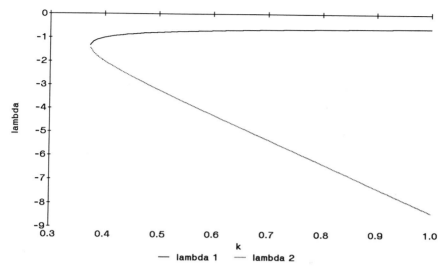

Figure 5.5 Graph of the eigenvalues λ_1 and λ_2 versus control parameter k in the docking problem: case $k \geq k_2$.

$$\lambda_1 = \frac{(1 - 10k) + \sqrt{(1 - 10k)^2 - 20k}}{2}$$

$$\lambda_2 = \frac{(1 - 10k) - \sqrt{(1 - 10k)^2 - 20k}}{2}$$

5.3   Phase Portraits

In Section 5.1 we introduced the eigenvalue test for stability in continuous-time dynamical systems. This test is based on the idea of a linear approximation in the neighborhood of an isolated equilibrium point. In this section we will show how this simple idea can be used to obtain a graphical description of the behavior of a dynamical system near an equilibrium point. This information can then be used along with a sketch of the vector field to obtain a graphical description of the dynamics over the entire state space, called the *phase portrait*. Phase portraits are important in the analysis of nonlinear dynamical systems because it is not possible in most cases to obtain exact analytical solutions. At the end of this section we also include a brief discussion of some similar techniques for discrete-time dynamical systems, again based on the idea of linear approximation.

Example 5.3   Consider the electrical circuit diagramed in Figure 5.6. The circuit consists of a capacitor, a resistor, and an inductor in a simple closed loop.

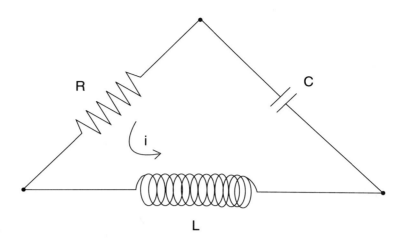

Figure 5.6 RLC circuit diagram for Example 5.3.

The effect of each component of the circuit is measured in terms of the relationship between current and voltage on that branch of the loop. An idealized physical model gives the relations

$$C\frac{dv_C}{dt} = i_C \text{ (capacitor)}$$

$$v_R = f(i_R) \text{ (resistor)}$$

$$L\frac{di_L}{dt} = v_L \text{ (inductor)},$$

where v_C represents the voltage across the capacitor, i_R represents the current through the resistor, and so on. The function $f(x)$ is called the *v–i characteristic* of the resistor. In the classical linear model of the RLC circuit we assume that $f(x) = Rx$, where R is the resistance. Kirchoff's current law states that the sum of the currents flowing into a node equals the sum of the currents flowing out. Kirchoff's voltage law states that the sum of the voltage drops along a closed loop must add up to zero. Determine the behavior of this circuit over time in the case where $L = 1$, $C = 1/3$, and $f(x) = x^3 + 4x$.

We will use the five-step method. The results of step 1 are summarized in Figure 5.7. Step 2 is to select a modeling approach. We will model this problem using a continuous-time dynamical system, which we will analyze by sketching the complete phase portrait.

Variables:
v_C = voltage across capacitor
i_C = current through capacitor
v_R = voltage across resistor
i_R = current through resistor
v_L = voltage across inductor
i_L = current through inductor

Assumptions:
$C \, dv_C/dt \quad = i_C$
$v_R \qquad\qquad = f(i_R)$
$L \, di_L/dt \quad = v_L$
$i_C \qquad\qquad = i_R = i_L$
$v_C + v_R + v_L = 0$
$L \qquad\qquad = 1$
$C \qquad\qquad = 1/3$
$f(x) \qquad\qquad = x^3 + 4x$

Objective:
Determine the behavior of all six variables over time

Figure 5.7 Step 1 of the RLC circuit problem.

Suppose that we are given a dynamical system $x' = F(x)$, where $x = (x_1, \ldots, x_n)$ and F has continuous first partial derivatives in the neighborhood of an equilibrium point x_0. Let A denote the matrix of first partial derivatives evaluated at the equilibrium point x_0 as defined by Eq. (1). We have stated previously that for x near x_0 the system $x' = F(x)$ behaves like the linear system $x' = A(x - x_0)$. Now we will be more specific.

The phase portrait of a continuous time dynamical system is simply a sketch of the state space showing a representative selection of solution curves. It is not hard to draw the phase portrait for a linear system (at least on $\mathbb{R}^2$), because we can always find an exact solution to a linear system of differential equations. Then we can just graph the solutions for a few initial conditions to get the phase portrait. We refer the reader to any textbook on differential equations for the details on how to solve linear systems of differential equations. For nonlinear systems, we can draw an approximate phase portrait in the neighborhood of each isolated equilibrium point by using the linear approximation.

A *homeomorphism* is a continuous function with a continuous inverse. The idea of a homeomorphism has to do with shapes and their generic properties. For example, consider a circle in the plane. The image of this circle under a homeomorphism

$$G : \mathbb{R}^2 \to \mathbb{R}^2$$

might be another circle or an ellipse, or even a square or a triangle. It could not be a line segment. This would violate continuity. It also could not be a figure eight, because this would violate the property that G must have an inverse (so it must be one-to-one). There is a theorem that states that if the eigenvalues of A all have nonzero real part, then there is a homeomorphism G that maps the phase portrait of the system $x' = Ax$ onto the phase portrait of $x' = F(x)$, with $G(0) = x_0$ (Hirsch, M. et al., (1974), p. 314). This theorem says that the phase portrait of $x' = F(x)$ around the point x_0 looks just like that of the linear system, except for some distortion. It would be as if we drew the phase portrait of the linear system on a sheet of rubber that we could stretch any way we like, but could not tear. This is a very powerful result. It means that we can get an actual picture (good enough for almost all practical purposes) of the behavior of a nonlinear dynamical system near each isolated equilibrium point just by analyzing its linear approximation. Then to finish up the phase portrait on the rest of the state space, we combine what we have learned about the behavior of solutions near the equilibrium points with the information contained in a sketch of the vector field.

Step 3 is to formulate the model. We begin by considering the state space. There are six state variables to begin with, but we can use Kirchoff's laws to reduce the number of degrees of freedom (the number of independent state variables) from six to two. Let $x_1 = i_R$, and notice that $x_1 = i_L = i_C$ as well. Let $x_2 = v_C$. Then we have

$$\frac{x_2'}{3} = x_1$$
$$v_R = x_1^3 + 4x_1$$
$$x_1' = v_L$$
$$x_2 + v_R + v_L = 0.$$

Substitute to obtain

$$\frac{x_2'}{3} = x_1$$
$$x_2 + x_1^3 + 4x_1 + x_1' = 0,$$

and then rearrange to get

$$x_1' = -x_1^3 - 4x_1 - x_2 \tag{15}$$
$$x_2' = 3x_1.$$

Now if we let $x = (x_1, x_2)$, then Eq. (15) can be written in the form $x' = F(x)$, where $F = (f_1, f_2)$ and

$$f_1(x_1, x_2) = -x_1^3 - 4x_1 - x_2 \tag{16}$$
$$f_2(x_1, x_2) = 3x_1.$$

This concludes step 3.

Step 4 is to solve the model. We will analyze the dynamical system in Eq. (15) by sketching the complete phase portrait. Figure 5.8 shows a sketch of the vector field for this dynamical system. Velocity vectors are horizontal on the curve $x_1 = 0$ where $x_2' = 0$, and vertical on the curve $x_2 = -x_1^3 - 4x_1$ where $x_1' = 0$. There is one equilibrium point, $(0, 0)$, at the intersection of these two curves. From the vector field it is difficult to tell whether the equilibrium is stable or unstable. To obtain more information we will analyze the linear system that approximates the behavior of Eq. (15) near the equilibrium $(0, 0)$.

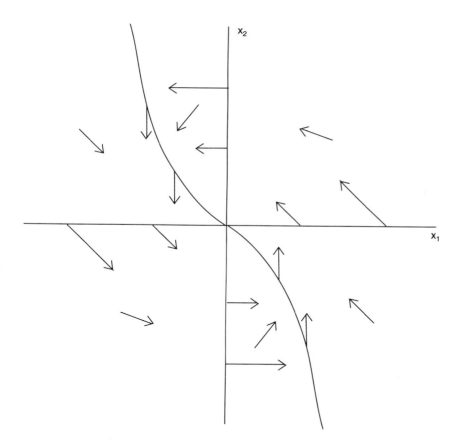

Figure 5.8 Graph of voltage x_2 versus current x_1 showing vector field for the RLC circuit problem of Example 5.3.

$$f_1(x_1, x_2) = -x_1^3 - 4x_1 - x_2$$
$$f_2(x_1, x_2) = 3x_1$$

Computing the partial derivatives from Eq. (16), we obtain

$$\frac{\partial f_1}{\partial x_1} = -3x_1^2 - 4$$

$$\frac{\partial f_1}{\partial x_2} = -1$$

$$\frac{\partial f_2}{\partial x_1} = 3$$

$$\frac{\partial f_2}{\partial x_2} = 0.$$

(17)

Evaluating the partial derivatives of Eq. (17) at the equilibrium point $(0, 0)$ and substituting back into Eq. (1) we obtain

$$A = \begin{pmatrix} -4 & -1 \\ 3 & 0 \end{pmatrix}.$$

The eigenvalues of this 2×2 matrix can be computed as the roots of the equation

$$\begin{vmatrix} \lambda + 4 & 1 \\ -3 & \lambda \end{vmatrix} = 0.$$

Evaluating the determinant, we obtain the equation

$$\lambda^2 + 4\lambda + 3 = 0,$$

and then we obtain

$$\lambda = -3, -1.$$

Since both eigenvalues are negative, the equilibrium is stable.

To obtain additional information we will solve the linear system $x' = Ax$. In this case we have

$$\begin{pmatrix} x_1' \\ x_2' \end{pmatrix} = \begin{pmatrix} -4 & -1 \\ 3 & 0 \end{pmatrix} \begin{pmatrix} x_1 \\ x_2 \end{pmatrix}.$$

(18)

We will solve the linear system of Eq. (18) by the method of eigenvalues and eigenvectors. We have already calculated the eigenvalues $\lambda = -3, -1$. To compute the eigenvector corresponding to the eigenvalue λ, we must find a nonzero solution to the equation

$$\begin{pmatrix} \lambda + 4 & 1 \\ -3 & \lambda \end{pmatrix} \begin{pmatrix} x_1 \\ x_2 \end{pmatrix} = \begin{pmatrix} 0 \\ 0 \end{pmatrix}.$$

For $\lambda = -3$ we have

$$\begin{pmatrix} 1 & 1 \\ -3 & -3 \end{pmatrix} \begin{pmatrix} x_1 \\ x_2 \end{pmatrix} = \begin{pmatrix} 0 \\ 0 \end{pmatrix},$$

from which we obtain

$$\begin{pmatrix} x_1 \\ x_2 \end{pmatrix} = \begin{pmatrix} -1 \\ 1 \end{pmatrix},$$

so that

$$\begin{pmatrix} -1 \\ 1 \end{pmatrix} e^{-3t}$$

is one solution to the linear system of Eq. (18). For $\lambda = -1$ we have

$$\begin{pmatrix} 3 & 1 \\ -3 & -1 \end{pmatrix} \begin{pmatrix} x_1 \\ x_2 \end{pmatrix} = \begin{pmatrix} 0 \\ 0 \end{pmatrix},$$

from which we obtain

$$\begin{pmatrix} x_1 \\ x_2 \end{pmatrix} = \begin{pmatrix} -1 \\ 3 \end{pmatrix},$$

so that

$$\begin{pmatrix} -1 \\ 3 \end{pmatrix} e^{-t}$$

is another solution to the linear system of Eq. (18). Then the general solution to Eq. (18) can be written in the form

$$\begin{pmatrix} x_1 \\ x_2 \end{pmatrix} = c_1 \begin{pmatrix} -1 \\ 1 \end{pmatrix} e^{-3t} + c_2 \begin{pmatrix} -1 \\ 3 \end{pmatrix} e^{-t}, \tag{19}$$

where c_1 and c_2 are arbitrary real constants. Figure 5.9 shows the phase portrait for the linear system of Eq. (18). This graph was obtained by sketching the solution curves in Eq. (19) for a few select values of the constants c_1 and c_2.

Figure 5.10 shows the complete phase portrait for the original nonlinear dynamical system in Eq. (15). This picture was obtained by combining the information in Figs. 5.8 and 5.9 and by using the fact that the phase portrait of the nonlinear system (15) is homeomorphic to the phase portrait of the linear system (18). In this example there is not much qualitative difference between the behavior of the linear and the nonlinear systems.

Step 5 is to answer the question. The question was to describe the behavior of the RLC circuit. The overall behavior can be described in terms of two quantities, the current through the resistor and the voltage drop across the capacitor. Regardless of the initial state of the circuit, both quantities eventually tend to zero. Furthermore, it is eventually true that either voltage is positive and current is negative, or vice versa. For a complete graphical description of the way that current and voltage behave over time, see Fig. 5.10, where x_1 represents current and x_2 represents voltage. The behavior of other quantities of interest can easily be described in terms of these two variables (see Fig. 5.7 for details). For example,

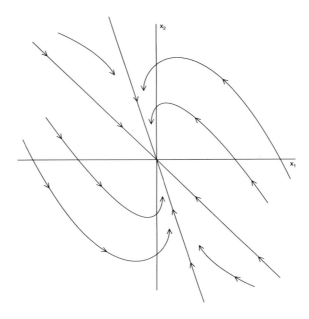

Figure 5.9 Graph of voltage x_2 versus current x_1 showing linear approximation to the phase portrait near $(0, 0)$ for the RLC circuit problem of Example 5.3.

the variable x_1 actually represents the current through any branch of the circuit loop.

Next we will perform a sensitivity analysis to determine the effect of small changes in our assumptions on our general conclusions. First let us consider the capacitance C. In our example we assumed that $C = 1/3$. Now we will generalize our model by letting C remain indeterminate. In this case we obtain the dynamical system

$$x_1' = -x_1^3 - 4x_1 - x_2$$
$$x_2' = \frac{x_1}{C} \tag{20}$$

in place of Eq. (15). Now we have

$$f_1(x_1, x_2) = -x_1^3 - 4x_1 - x_2$$
$$f_2(x_1, x_2) = \frac{x_1}{C}. \tag{21}$$

For values of C near $1/3$, the vector field for Eq. (21) is essentially the same as in Fig. 5.8. Velocity vectors are still horizontal on the curve $x_1 = 0$, and vertical on the curve $x_2 = -x_1^3 - 4x_1$. There is still one equilibrium point, $(0, 0)$, at the intersection of these two curves.

Computing the partial derivatives from Eq. (21), we obtain

$$
\begin{aligned}
\frac{\partial f_1}{\partial x_1} &= -3x_1^2 - 4 \\
\frac{\partial f_1}{\partial x_2} &= -1 \\
\frac{\partial f_2}{\partial x_1} &= \frac{1}{C} \\
\frac{\partial f_2}{\partial x_2} &= 0.
\end{aligned}
\tag{22}
$$

Evaluating the partial derivatives in Eq. (22) at the equilibrium point $(0, 0)$ and

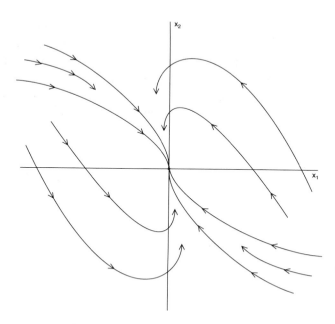

Figure 5.10 Graph of voltage x_2 versus current x_1 showing complete phase portrait for the RLC circuit problem of Example 5.3.

substituting back into Eq. (1), we obtain

$$A = \begin{pmatrix} -4 & -1 \\ 1/C & 0 \end{pmatrix}.$$

The eigenvalues of this matrix can be computed as the roots of the equation

$$\begin{vmatrix} \lambda + 4 & 1 \\ -1/C & \lambda \end{vmatrix} = 0.$$

Evaluating the determinant, we obtain the equation

$$\lambda^2 + 4\lambda + 1/C = 0.$$

The eigenvalues are

$$\lambda = -2 \pm \sqrt{4 - \frac{1}{C}}.$$

If $C > 1/4$, we have two distinct real negative eigenvalues, and so the equilibrium is stable. In this case the general solution to the linear system is

$$\begin{pmatrix} x_1 \\ x_2 \end{pmatrix} = c_1 \begin{pmatrix} -1 \\ 2 + \alpha \end{pmatrix} e^{(-2+\alpha)t} + c_2 \begin{pmatrix} -1 \\ 2 - \alpha \end{pmatrix} e^{(-2-\alpha)t}, \tag{23}$$

where $\alpha^2 = 4 - 1/C$. The phase portrait of the linear system is about the same as Fig. 5.9 except that the slope of the straight-line solutions varies with C. Then for values of C greater than $1/4$, the phase portrait for the original nonlinear system is a lot like the one shown in Fig. 5.10. We conclude that our general conclusions about this RLC circuit are not sensitive to the exact value of C as long as $C > 1/4$. A similar result may be expected for the inductance L. Generally speaking, the important characteristics of our solution (e.g., eigenvectors) depend continuously on these parameters.

Next we consider the question of robustness. We assumed that the RLC circuit had v–i characteristic $f(x) = x^3 + 4x$. Suppose more generally that $f(0) = 0$ and that f is strictly increasing. Now the dynamical system equations are

$$\begin{aligned} x_1' &= -f(x_1) - x_2 \\ x_2' &= 3x_1. \end{aligned} \tag{24}$$

Now we have

$$\begin{aligned} f_1(x_1, x_2) &= -f(x_1) - x_2 \\ f_2(x_1, x_2) &= 3x_1. \end{aligned} \tag{25}$$

Let $R = f'(0)$. The linear approximation uses

$$A = \begin{pmatrix} -R & -1 \\ 3 & 0 \end{pmatrix},$$

thus the eigenvalues are the roots to the equation

$$
\begin{vmatrix} \lambda + R & 1 \\ 3 & \lambda \end{vmatrix} = 0.
$$

We compute that

$$
\lambda = \frac{-R \pm \sqrt{R^2 - 12}}{2}.
$$

As long as $R > \sqrt{12}$ we have two distinct real negative eigenvalues, and the behavior of the linear system is as depicted in Fig. 5.9. Furthermore, the behavior of the original nonlinear system cannot be too different from Fig. 5.10. We conclude that our model of this RLC circuit is robust with regard to our assumptions about the form of the v–i characteristic.

Example 5.4 Consider the nonlinear RLC circuit with $L = 1$, $C = 1$, and v–i characteristic $f(x) = x^3 - x$. Determine the behavior of this circuit over time.

The modeling process is of course the same as for the previous example. Letting $x_1 = i_R$ and $x_2 = v_C$, we obtain the dynamical system

$$
\begin{aligned}
x_1' &= x_1 - x_1^3 - x_2 \\
x_2' &= x_1.
\end{aligned}
\tag{26}
$$

See Figure 5.11 for a sketch of the vector field. The velocity vectors are vertical on the curve $x_2 = x_1 - x_1^3$ and horizontal on the x_2 axis. The only equilibrium is the origin, $(0, 0)$. It is hard to tell from the vector field whether or not the origin is a stable equilibrium.

The matrix of partial derivatives is

$$
A = \begin{pmatrix} 1 - 3x_1^2 & -1 \\ 1 & 0 \end{pmatrix}.
$$

Evaluate at $x_1 = 0$, $x_2 = 0$ to obtain the linear system

$$
\begin{pmatrix} x_1' \\ x_2' \end{pmatrix} = \begin{pmatrix} 1 & -1 \\ 1 & 0 \end{pmatrix} \begin{pmatrix} x_1 \\ x_2 \end{pmatrix},
$$

which approximates the behavior of our nonlinear system near the origin. To obtain the eigenvalues we must solve

$$
\begin{vmatrix} \lambda - 1 & 1 \\ -1 & \lambda - 0 \end{vmatrix} = 0,
$$

or $\lambda^2 - \lambda + 1 = 0$. The eigenvalues are

$$
\lambda = 1/2 \pm i\sqrt{3}/2.
$$

Since the real part of every eigenvalue is positive, the origin is an unstable equilibrium.

To obtain more information we will solve the linear system. To find an eigenvector belonging to

$$\lambda = 1/2 + i\sqrt{3}/2,$$

we solve

$$\begin{pmatrix} -1/2 + i\sqrt{3}/2 & 1 \\ -1 & 1/2 + i\sqrt{3}/2 \end{pmatrix} \begin{pmatrix} x_1 \\ x_2 \end{pmatrix} = \begin{pmatrix} 0 \\ 0 \end{pmatrix},$$

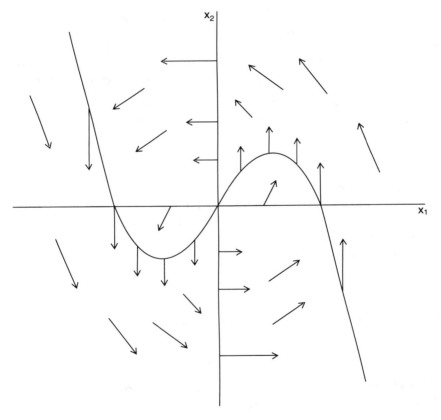

Figure 5.11 Graph of voltage x_2 versus current x_1 showing vector field for the RLC circuit problem of Example 5.4.

$$f_1(x_1, x_2) = x_1 - x_1^3 - x_2$$
$$f_2(x_1, x_2) = x_1$$

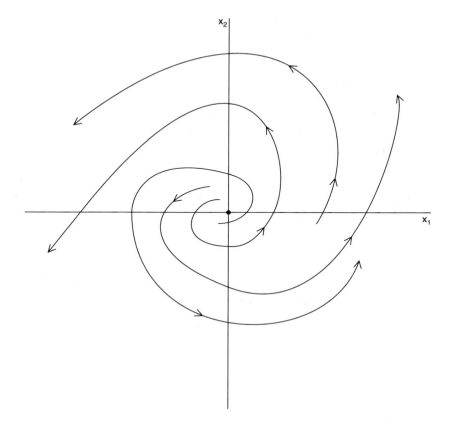

Figure 5.12 Graph of voltage x_2 versus current x_1 showing linear approximation to the phase portrait near $(0, 0)$ for the RLC circuit problem of Example 5.4.

and we obtain

$$x_1 = 2, \ x_2 = 1 - i\sqrt{3}.$$

Then we have the complex solution

$$\begin{pmatrix} x_1 \\ x_2 \end{pmatrix} = \begin{pmatrix} 2 \\ 1 - i\sqrt{3} \end{pmatrix} e^{(\frac{1}{2} + i\frac{\sqrt{3}}{2})t}.$$

Taking real and imaginary parts yields two linearly independent real solutions

$$x_1(t) = 2e^{t/2} \cos(t\sqrt{3}/2)$$
$$x_2(t) = e^{t/2} \cos(t\sqrt{3}/2) + \sqrt{3} \, e^{t/2} \sin(t\sqrt{3}/2)$$

and

$$x_1(t) = 2e^{t/2}\sin(t\sqrt{3}/2)$$
$$x_2(t) = e^{t/2}\sin(t\sqrt{3}/2) - \sqrt{3}\,e^{t/2}\cos(t\sqrt{3}/2).$$

The general solution is obtained by taking linear combinations of these two real solutions. Figure 5.12 shows the phase portrait for this linear system. The phase portrait for the nonlinear system in the neighborhood of the origin looks the same, with some distortions. The solution curves near (0, 0) spiral outward moving counterclockwise.

It is apparent from the sketch of the vector field in Fig. 5.11 that solution curves far away from the origin also rotate counterclockwise. You might be led to conjecture that all of the solution curves simply spiral out to infinity; however, this is not the case. In Section 6.3 we will explore the behavior of the dynamical system in Eq. (26), using computational methods. We will wait until then to draw the complete phase portrait.

Before we leave the subject of linear approximation techniques, we should point out a few facts about discrete-time dynamical systems. Suppose we have a discrete-time dynamical system

$$\Delta x = F(x),$$

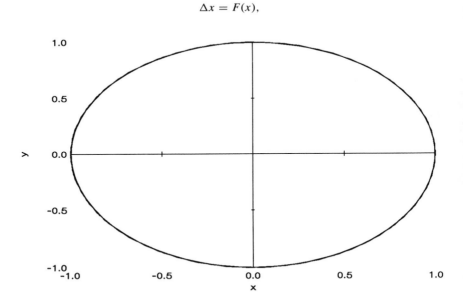

Figure 5.13 Dynamics of the docking problem, initial condition.

$$S = \{(x,\ y) : x^2 + y^2 = 1\}$$

where $x = (x_1, \ldots, x_n)$, and let

$$G(x) = x + F(x)$$

denote the iteration function. At an equilibrium point x_0 we have $G(x_0) = x_0$. In Section 5.2 we used the approximation

$$G(x) \approx A(x - x_0)$$

for values of x near x_0, where A is the matrix of partial derivatives evaluated at $x = x_0$ as defined by Eq. (12). One way to obtain a graphical picture of the iteration function $G(x)$ is to draw the image sets

$$G(S) = \{G(x) : x \in S\}$$

for various sets

$$S = \{x : |x - x_0| = r\}.$$

In dimension $n = 2$ the set S is a circle, and in dimension $n = 3$ it is a sphere. It is possible to show that as long as the matrix A is nonsingular there is a diffeomorphism $H(x)$ that maps the image sets $A(S)$ onto $G(S)$ in a neighborhood of the point x_0. If a point x lies inside of S, then $G(x)$ will be inside of $G(S)$. This allows a graphical interpretation of the dynamics. Figures 5.13 through 5.15 illustrate the dynamics of the docking problem

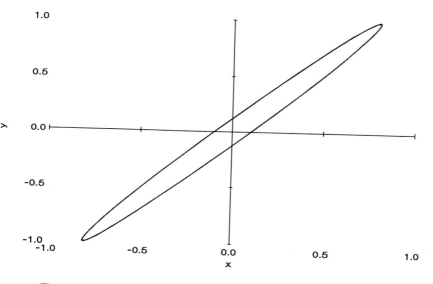

Figure 5.14 Dynamics of the docking problem, after one iteration.

$A(S)$

from Example 5.2. In this case $G(S) = A(S)$ since G is linear. Starting at a state on (or inside) the set S, shown in Fig. 5.13, the next state will be on (or inside, respectively) the set $A(S)$ shown in Fig. 5.14, and then the next state will be on (or inside, respectively) the set $A^2(S) = A(A(S))$ shown in Fig. 5.15. As $n \to \infty$, the set $A^n(S)$ gradually shrinks in toward the origin.

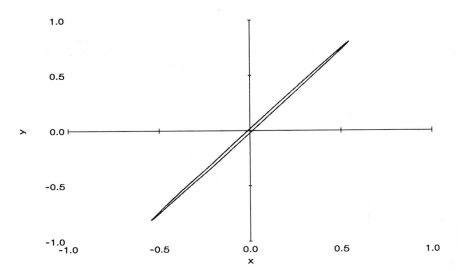

Figure 5.15 Dynamics of the docking problem, after two iterations.

$$A^2(S)$$

5.4 Exercises

1. Reconsider Exercise 4 of Chapter 4.
   (a) Sketch the vector field for this model. Determine the location of each equilibrium in the state space. Can you tell from the vector field which of the equilibria are stable?
   (b) Use the eigenvalue method to test the stability of each equilibrium in the state space.
   (c) For each equilibrium point, determine the linear system that approximates the behavior of the original dynamical system in the neighborhood

of the equilibrium point, and then sketch the phase portrait for the linear system.

(d) Sketch the complete phase portrait for this model, using the results of parts (a) and (c) above.

(e) Given the current estimates of 5,000 blue whales and 70,000 fin whales, what does this model predict about the future of the two species?

2. Reconsider Exercise 5 of Chapter 4.

   (a) Sketch the vector field for this model. Determine the location of each equilibrium in the state space.

   (b) Use the eigenvalue method to test the stability of each equilibrium in the state space.

   (c) For each equilibrium point, determine the linear system that approximates the behavior of the original dynamical system in the neighborhood of the equilibrium point, and then sketch the phase portrait for the linear system.

   (d) Sketch the complete phase portrait for this model, using the results of parts (a) and (c) above.

   (e) Given the current estimates of 5,000 blue whales and 70,000 fin whales, what does this model predict about the future of the two species?

3. Reconsider Exercise 6 of Chapter 4. Assume that the catchability coefficient is $q = 10^{-5}$ and that the level of effort is $E = 3,000$ boat-days per year.

   (a) Sketch the vector field for this model. Determine the location of each equilibrium in the state space.

   (b) Use the eigenvalue method to test the stability of each equilibrium in the state space.

   (c) For each equilibrium point, determine the linear system that approximates the behavior of the original dynamical system in the neighborhood of the equilibrium point, and then sketch the phase portrait for the linear system.

   (d) Sketch the complete phase portrait for this model, using the results of parts (a) and (c) above.

   (e) Given the current estimates of 5,000 blue whales and 70,000 fin whales, what does this model predict about the future of the two species?

4. Repeat Exercise 3 above, but now assume a level of effort of $E = 6,000$ boat-days per year.

5. Reconsider Exercise 7 of Chapter 4.

   (a) Sketch the vector field for this model. Determine the location of each equilibrium in the state space.

(b) Use the eigenvalue method to test the stability of each equilibrium in the state space.

(c) For each equilibrium point, determine the linear system that approximates the behavior of the original dynamical system in the neighborhood of the equilibrium point, and then sketch the phase portrait for the linear system.

(d) Sketch the complete phase portrait for this model, using the results of parts (a) and (c) above.

(e) Suppose that an ecological disaster suddenly kills off 80% of the krill in the area, leaving 150,000 blue whales and only 100 tons/acre of krill. What does our model predict about the future of the whales and the krill?

6. Reconsider Exercise 9 of Chapter 4.

(a) Use an eigenvalue method to determine the stability of the (P, Q) equilibrium, assuming a continuous-time model.

(b) Repeat part (a), assuming a discrete-time model. To what do you attribute the difference in your results?

7. Reconsider the tree problem of Example 5.1. Assume $t = 1/2$.

(a) Sketch the vector field for this model. Indicate the location of each equilibrium in the state space.

(b) For each equilibrium point, determine the linear system that approximates the behavior of the original dynamical system in the neighborhood of the equilibrium point, and then sketch the phase portrait for the linear system.

(c) Sketch the complete phase portrait for this model, using the results of parts (a) and (b) above.

(d) Suppose that a small number of hardwood trees is introduced into a mature stand of softwood trees. What does our model predict about the future of this forest?

8. Reconsider the tree problem of Example 5.1, but now suppose that the strength of the competition factor is too great to allow the coexistence of both hardwood and softwood trees. Assume $t = 3/4$.

(a) Sketch the vector field for this model. Indicate the location of each equilibrium in the state space.

(b) For each equilibrium point, determine the linear system that approximates the behavior of the original dynamical system in the neighborhood of the equilibrium point, and then sketch the phase portrait for the linear system.

(c) Sketch the complete phase portrait for this model, using the results of parts (a) and (b) above.

(d) Suppose that a small number of hardwood trees is introduced into a mature stand of softwood trees. What does our model predict about the future of this forest?

9. Reconsider the RLC circuit problem of Example 5.3, and perform a sensitivity analysis on the parameter L, which gives the inductance of the capacitor. Assume that $L > 0$.

(a) Describe the vector field for the general case $L > 0$.

(b) Determine the range of L for which the equilibrium $(0, 0)$ remains stable.

(c) Draw the phase portrait for the linear system in the case where there are two real eigenvalues.

(d) Use the results of parts (a) and (c) to draw the complete phase portrait. Comment on the sensitivity of our conclusions in Section 5.3 to the actual value of the parameter L.

10. Reconsider the RLC circuit problem of Example 5.3, but now suppose that the capacitance $C = 1/5$.

(a) Sketch the vector field for this model.

(b) Use the eigenvalue method to test the stability of the equilibrium at the origin.

(c) Determine the linear system that approximates the behavior of the original dynamical system in the neighborhood of the origin, and then sketch the phase portrait for the linear system.

(d) Sketch the complete phase portrait for this model, using the results of parts (a) and (c) above. How does the behavior of the RLC circuit change when the capacitance C is lowered?

11. (Continuation of Exercise 10) Reconsider the RLC circuit problem of Example 5.3, and now consider the effect of varying the capacitance C over the entire range $0 < C < \infty$.

(a) Sketch the phase portrait of the linear system that approximates the behavior of the RLC circuit in the neighborhood of the origin in the case $0 < C < 1/4$. Compare with the case $C > 1/4$ that was done in the text.

(b) Draw the complete phase portrait for the RLC circuit for the case $0 < C < 1/4$. Describe the changes that occur in the phase portrait as we transition between the two cases $0 < C < 1/4$ and $C > 1/4$.

(c) Draw the phase portrait of the linear system in the case $C = 1/4$ where there is only one eigenvalue. Sketch the phase portrait of the nonlinear system in this case. Explain how the phase portrait in this case represents an intermediate step between the case of two real distinct eigenvalues ($C > 1/4$) and one pair of complex conjugate eigenvalues ($0 < C < 1/4$).

(d) Reconsider the description of circuit behavior given in step 5 of Example 5.3 in the text. Describe in plain English the behavior of the RLC circuit in the more general case $C > 0$.

12. Reconsider the space docking problem of Example 5.2.

(a) Draw the vector field for this problem.

(b) Find the eigenvectors associated with the eigenvalues

$$\lambda = \frac{4 \pm \sqrt{6}}{10},$$

which were calculated in the text. Draw these eigenvectors on the picture from part (a). What do you notice about the vector field at these points?

(c) Calculate the rate at which closing velocity decreases (% per minute) if we start at an eigenvector.

(d) Generally speaking, what can you say about the rate of decrease in closing velocity for an arbitrary initial condition? [Hint: Any (x_1, x_2) initial condition is a linear combination of the two eigenvectors found in part (b).]

13. (Continuation of Exercise 12) Reconsider the space docking problem of Example 5.2.

(a) As in Exercise 12, calculate the rate at which closing velocity decreases (% per minute) as a function of the control parameter k over the range of values $0 < k \leq 0.0268$.

(b) Find the value of k in part (a) that maximizes the rate of decrease in closing velocity.

(c) Explain the significance of your results in part (b) in terms of the efficiency of the control parameter k.

(d) Explain the problem with extending the approach used in this problem to find the most efficient value of k over the entire interval $0 < k < 0.2$, over which we have a stable control procedure.

14. Reconsider the space docking problem of Example 5.2, but now approximate the discrete-time dynamical system

$$\Delta x_1 = -kwx_1 - kcx_2$$

$$\Delta x_2 = x_1 - x_2$$

by its continuous-time analogue

$$\frac{dx_1}{dt} = -kwx_1 - kcx_2$$

$$\frac{dx_2}{dt} = x_1 - x_2.$$

(a) Show that the continuous-time model has a stable equilibrium at (0, 0). Assume $w = 10$, $c = 5$, and $k = 0.02$.

(b) Solve the continuous model, using the method of eigenvalues and eigenvectors.

(c) Draw the complete phase portrait for this model.

(d) Comment on any differences between the behavior of the discrete and continuous models.

15. (Continuation of Exercise 14) Reconsider the docking problem of Example 5.2. As in Exercise 14, replace the discrete-time model by its continuous-time analogue.

(a) Assume $w = 10$ and $c = 5$. For what values of k does the continuous model have a stable equilibrium at (0, 0)?

(b) Solve the continuous model using the method of eigenvectors and eigenvalues.

(c) Draw the complete phase portrait for this model. How does the phase portrait depend on k?

(d) Comment on any differences between the continuous and discrete models.

16. Reconsider the tree problem of Example 5.1.

(a) Can both types of trees coexist in stable equilibrium? Assume $b_i = a_i/2$. Use the five-step method, and model as a discrete-time dynamical system with a time step of one year.

(b) Use the eigenvalue test for discrete-time dynamical systems to check the stability of the equilibrium you found in part (a).

(c) Perform a sensitivity analysis on the parameter t, where $b_i = ta_i$. Determine the range of $0 < t < 0.6$ for which the equilibrium found in part (a) is stable.

(d) Comment on any differences in results between the discrete-time and continuous-time models. As a practical matter, does it make any difference which we choose?

Further Reading

Beltrami, E. (1987). *Mathematics for Dynamic Modeling*, Academic Press, Orlando, Florida.

Frauenthal, J. (1979). *Introduction to Population Modeling*, UMAP Monograph.

Hirsch, M. and Smale, S. (1974). *Differential Equations, Dynamical Systems, and Linear Algebra*, Academic Press, New York.

Keller, M. *Electrical Circuits and Applications of Matrix Methods: Analysis of Linear Circuits*, UMAP modules 108 and 112.

Press, W., Flannery, B., Teukolsky, S. and Vetterling, W. (1987). *Numerical Recipes*, Cambridge University Press, New York.

Rescigno, A. and Richardson, I., The Struggle for Life I, Two Species, *Bulletin of Mathematical Biophysics*, Vol. 29, pp. 377-388.

Smale, S. (1972).   On the Mathematical Foundations of Circuit Theory, *Journal of Differential Geometry*, Vol. 7, pp. 193-210.

Wilde, C. *The Contraction Mapping Principle*, UMAP module 326.

Chapter Six

Simulation of Dynamic Models

The technique of simulation has become the most important and popular method of analysis for dynamic models. Exact solution methods such as those taught in introductory courses in differential equations are of limited scope. The fact is that we do not know how to solve very many differential equations. The qualitative methods introduced in the two preceding chapters are more widely applicable, but for some problems we need a quantitative answer and a high degree of accuracy. Simulation methods provide both. Almost any dynamic model that will occur in practical applications can be simulated to a reasonable degree of accuracy. Furthermore, simulation techniques are very flexible. It is not hard to introduce more complex features such as time delays and stochastic elements, which are difficult to treat analytically.

The principal drawback of simulation comes in the area of sensitivity analysis. Without recourse to an analytic formula, the only way to test sensitivity to a particular parameter is to repeat the entire simulation for several values and then interpolate. This can be very expensive and time-consuming if there are several parameters to test. Even so, simulation is the method of choice for many problems. If we cannot solve analytically, and if we need quantitative solutions, then we really have no alternative but to simulate.

6.1 Introduction to Simulation

There are essentially two ways to approach the analysis of a dynamic system model. The analytic approach attempts to predict what will happen according to the model in a variety of circumstances. In the simulation approach, we build the model, turn it on, and find out.

Example 6.1   Two forces, which we will call red (R) and blue (B), are engaged in battle. In this conventional battle, attrition is due to direct fire (infantry) and area fire (artillery). The attrition rate due to direct fire is assumed proportional to the number of enemy infantry. The attrition rate due to artillery depends on both the amount of enemy artillery and the density of friendly troops. Red has amassed five divisions to attack a blue force of two divisions. Blue has the advantage of defense, and superior weapon effectiveness besides. How much more effective does blue have to be in order to prevail in battle?

We will use the five-step method. The results of step 1 are summarized in Figure 6.1. We have assumed that the attrition rate due to area fire is directly proportional to the product of enemy force level and friendly force level. It seems reasonable to assume at this stage that force level is proportional to force density. And since we have no information about the number of artillery versus infantry units, for this analysis we simply assume that artillery and infantry units are attrited in proportion to their numbers. Hence, the number of remaining artillery or infantry units on each side is assumed to remain in proportion to the total number of units.

Next is step 2. We will use a discrete-time dynamical system model, which we will solve by simulation. Figure 6.2 gives an algorithm for simulating a discrete-time dynamical system in two variables:

$$\Delta x_1 = f_1(x_1, x_2)$$
$$\Delta x_2 = f_2(x_1, x_2). \tag{1}$$

Step 3 is next. We will model the war problem as a discrete-time dynamical system with two state variables: $x_1 = R$, the number of red force units; and $x_2 = B$, the number of blue force units. The difference equations are

$$\Delta x_1 = -a_1 x_2 - b_1 x_1 x_2$$
$$\Delta x_2 = -a_2 x_1 - b_2 x_1 x_2. \tag{2}$$

We will start with $x_1(0) = 5$ and $x_2(0) = 2$ divisions of troops. We will use a time step of $\Delta t = 1$ hour. We will also need numerical values for a_i and b_i in order to run the simulation program. Unfortunately we have been given no idea

Variables: R = number of red units (divisions)

B = number of blue units (divisions)

D_R = red attrition rate due to direct fire (units/hour)

D_B = blue attrition rate due to direct fire (units/hour)

I_R = red attrition rate due to indirect fire (units/hour)

I_B = blue attrition rate due to indirect fire (units/hour)

Assumptions: $D_R = a_1 B$

$D_B = a_2 R$

$I_R = b_1 R B$

$I_B = b_2 R B$

$R \geq 0, \ B \geq 0$

$R(0) = 5, \ B(0) = 2$

$a_1, \ a_2, \ b_1, \ b_2$ are positive reals

$a_1 > a_2, \ b_1 > b_2$

Objective: Determine the conditions under which

$R \to 0$ before $B \to 0$

Figure 6.1 Results of step 1 of the war problem.

what they are supposed to be. We will have to make an educated guess. Suppose that a typical conventional battle lasts about 5 days and that engagement takes place about 12 hours per day. That means that one force is depleted in about 60 hours of battle. If a force were to be depleted by 5% per hour for 60 hours, the fraction remaining would be $(0.95)^{60} = 0.05$, which looks about right. We will assume that $a_2 = 0.05$. Since area fire is not generally as effective as direct fire in terms of attrition, we will assume $b_2 = 0.005$. (Recall that b_i is multiplied by both x_1 and x_2, which is why we made it so small.) Now blue is supposed to have greater weapon effectiveness than red, so we should have $a_1 > a_2$ and $b_1 > b_2$. Let us assume that $a_1 = \lambda a_2$ and $b_1 = \lambda b_2$ for some $\lambda > 1$. The analysis objective is to determine the smallest λ that will make $x_1 \to 0$ before $x_2 \to 0$. Now the difference equations are

$$\Delta x_1 = -\lambda(0.05)x_2 - \lambda(0.005)x_1 x_2$$
$$\Delta x_2 = -0.05x_1 - 0.005x_1 x_2. \tag{3}$$

In step 4 we will solve the problem by running the simulation program for several values of λ. We begin by exercising the model for $\lambda = 1, 1.5, 2, 3,$ and 5.

Algorithm: Discrete-time simulation

Variables: $x_1(n)$ = first state variable at time n
                   $x_2(n)$ = second state variable at time n
                   N = number of time steps

Input: $x_1(0),\ x_2(0),\ N$

Process: Begin
                 for $n = 1$ to N do
                     Begin
                     $x_1(n) \leftarrow x_1(n-1) + f_1(x_1(n-1),\ x_2(n-1))$
                     $x_2(n) \leftarrow x_2(n-1) + f_2(x_1(n-1),\ x_2(n-1))$
                     End
                 End

Output: $x_1(1),\ \ldots,\ x_1(N)$
                $x_2(1),\ \ldots,\ x_2(N)$

Figure 6.2 Pseudocode for discrete-time simulation.

This should give us a good idea of how large λ needs to be, and it allows us to check our simulation against our intuitive grasp of the situation. For example, we should check that the larger λ is, the better blue does.

The results of this first batch of model runs are shown in Figures 6.3 through 6.7. A summary of our findings is contained in Table I. For each run we have recorded the value of λ, the duration of the battle, the identity of the winner, and the number of units remaining on the winning side. We decided to run the simulation for up to 14 days of combat (or $N = 168$ hours). The duration of battle is defined to be the number of hours of actual combat (there are 12 hours of fighting per day) until one of the variables x_1 or x_2 becomes zero or negative. If both sides survive 168 hours of combat, we call it a draw.

It does not look good for blue. Even with a 5:1 edge in weapon effectiveness, blue will lose the battle. We decided to make a few more model runs to find out just how big λ would have to be for blue to win. At $\lambda = 6.0$, the blue side won after 13 hours of battle, with 0.6 units remaining (see Figure 6.8). A few more model runs, bisecting the interval $5.0 \leq \lambda \leq 6.0$, yielded a lower bound of $\lambda = 5.4$ for blue to win. At $\lambda = 5.3$, red was the winner.

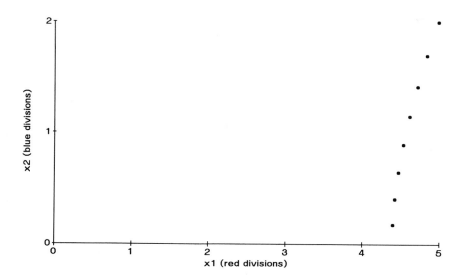

Figure 6.3 Graph of blue divisions x_2 versus red divisions x_1 for the war problem: case $\lambda = 1.0$.

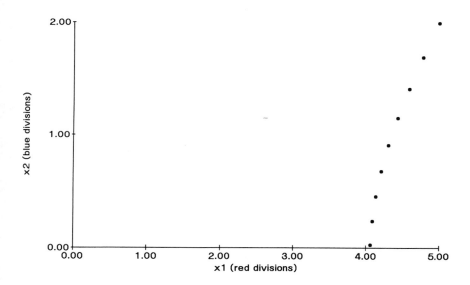

Figure 6.4 Graph of blue divisions x_2 versus red divisions x_1 for the war problem: case $\lambda = 1.5$.

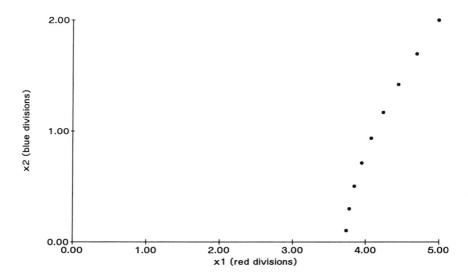

Figure 6.5 Graph of blue divisions x_2 versus red divisions x_1 for the war problem: case $\lambda = 2.0$.

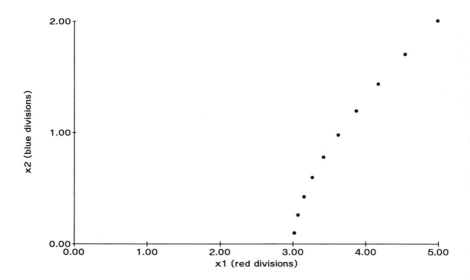

Figure 6.6 Graph of blue divisions x_2 versus red divisions x_1 for the war problem: case $\lambda = 3.0$.

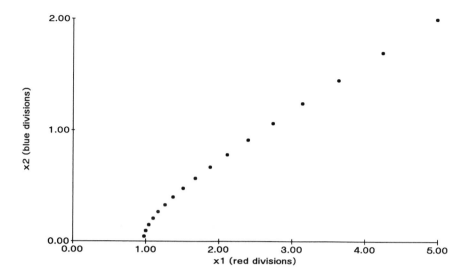

Figure 6.7 Graph of blue divisions x_2 versus red divisions x_1 for the war problem: case $\lambda = 5.0$.

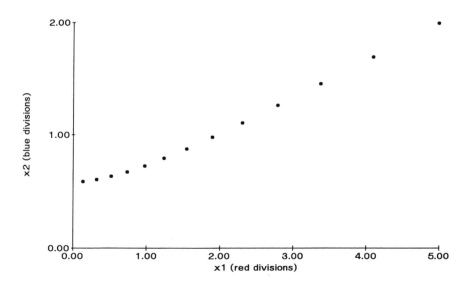

Figure 6.8 Graph of blue divisions x_2 versus red divisions x_1 for the war problem: case $\lambda = 6.0$.

Table I.   Summary of simulation results for the war problem.

Advantage (λ)	Hours of combat	Winning side	Remaining forces
1.0	8	red	4.4
1.5	9	red	4.1
2.0	9	red	3.7
3.0	10	red	3.0
5.0	17	red	1.0

Finally, we need to summarize our results. We simulated an engagement between an attacking red force of five divisions and a defending blue force of two divisions. We assumed that the two forces would engage and continue to fight until there emerged a clear victor. We wanted to investigate the extent to which a greater weapon effectiveness (kill rate) could offset a 5:2 numerical disadvantage. We simulated a number of battles for different ratios of weapon effectiveness. We found that blue would need at least a 5.4:1 advantage in weapon effectiveness to defend successfully against a numerically superior red force of 5 divisions.

Having finished the five-step method and answered the question stated in step 1, we need to perform a sensitivity analysis. This is particularly important in a problem such as this one, where most of our data came from sheer guesswork. We will begin by investigating the relationship between the magnitude of the attrition coefficients and the outcome of the battle. We had assumed that $a_2 = 0.05$, $b_2 = a_2/10$, $a_1 = \lambda a_2$, and $b_1 = \lambda b_2$. We will now vary a_2, keeping the same relative relationship between it and the other variables.

We will investigate the dependence of λ_{min} on a_2, where λ_{min} is defined to be the smallest value of λ for which blue wins. This requires making a number of model runs for each value of a_2. It turns out that there is no need to tabulate these results, because in every case we checked (from $a_2 = 0.01$ to $a_2 = 0.10$) we found $\lambda_{min} = 5.4$, the same as our baseline case ($a_2 = 0.05$). There is apparently no sensitivity to the magnitude of the attrition coefficients.

Several more kinds of sensitivity analysis are possible, and the process probably should go on as long as time permits, curiosity persists, and the pressures of other obligations do not intrude. We were curious about the relationship between λ_{min} and the numerical superiority ratio of red versus blue, currently assumed to be 5:2. To study this, we returned to the baseline case, $a_2 = 0.05$, and made several model runs to determine λ_{min} for several values of the initial red force strength

Table II.   Summary of simulation results showing the
effect of force ratio for the war problem.

Force ratio (red:blue)	Advantage required (λ_{min})
8:2	11.8
7:2	9.5
6:2	7.3
5:2	5.4
4:2	3.6
3:2	2.2
2:2	1.1

x_1, keeping $x_2 = 2$ fixed. The results of this model excursion are tabulated in Table II. The case $x_1 = 2$ was run as a check. We found $\lambda_{min} = 1.1$ in this case, because the case $\lambda = 1$ produced a draw.

6.2   Continuous-Time Models

In this section we discuss the fundamentals of simulating continuous-time dynamical systems. The methods presented here are simple and usually effective. The basic idea is to use the approximation

$$dx/dt \approx \Delta x/\Delta t$$

to replace our continuous-time model (differential equations) by a discrete-time model (difference equations). Then we can use the simulation methods we introduced in the preceding section.

Example 6.2   Reconsider the whale problem of Example 4.2. We know now that starting at the current population levels of $B = 5,000$, $F = 70,000$, and assuming a competition coefficient of $\alpha < 1.25 \times 10^{-7}$, both populations of whales will eventually grow back to their natural levels in the absence of any further harvesting. How long will this take?

We will use the five-step method. Step 1 is the same as before (see Fig. 4.3), except that now the objective is to determine how long it takes to get to the equilibrium starting from $B = 5,000$, $F = 70,000$.

Step 2 is to select the modeling approach. We have an analysis question that seems to require a quantitative method. The graphical methods of Chapter 4 tell us what will happen, but not how long it will take. The analytical methods reviewed in Chapter 5 are local in nature. We need a global method here. The best thing would be to solve the differential equations, but we don't know how. We will use a simulation—this seems to be the only choice we have.

There is some question as to whether we want to adopt a discrete-time or a continuous-time model. Let us consider, more generally, the case of a dynamic model in n variables, $x = (x_1, \ldots, x_n)$, where we are given the rates of change $F = (f_1, \ldots, f_n)$ for each of the variables $x_1, \ldots, x_n$ but we have not yet decided whether to model the system in discrete-time or continuous-time. The discrete-time model looks like

$$\Delta x_1 = f_1(x_1, \ldots, x_n)$$

$$\vdots \tag{4}$$

$$\Delta x_n = f_n(x_1, \ldots, x_n),$$

where Δx_i represents the change in x_i over 1 unit of time ($\Delta t = 1$). The units of time are already specified. The method for simulating such a system was discussed in the previous section.

If we decided on a continuous-time model instead, we would have

$$\frac{dx_1}{dt} = f_1(x_1, \ldots, x_n)$$

$$\vdots \tag{5}$$

$$\frac{dx_n}{dt} = f_n(x_1, \ldots, x_n),$$

which we would still need to figure out how to simulate. We certainly can't expect the computer to calculate $x(t)$ for every value of t. That would take an infinite amount of time to get nowhere. Instead we must calculate $x(t)$ at a finite number of points in time. In other words, we must replace the continuous-time model by a discrete-time model in order to simulate it. What would the discrete-time approximation to this continuous-time model look like? If we use a time step of $\Delta t = 1$ unit, it will be exactly the same as the discrete-time model we could have chosen in the first place. Hence, unless there is something wrong with choosing $\Delta t = 1$, we don't have to choose between discrete and continuous. Then we are done with step 2.

Step 3 is to formulate the model. As in Chapter 4 we let $x_1 = B$ and $x_2 = F$ represent the population levels of each species. The dynamical system equations are

$$\frac{dx_1}{dt} = .05x_1(1 - x_1/150,000) - \alpha x_1 x_2$$
$$\frac{dx_2}{dt} = .08x_2(1 - x_2/400,000) - \alpha x_1 x_2$$

(6)

on the state space $x_1 \geq 0$, $x_2 \geq 0$. In order to simulate this model we will begin by transforming to a set of difference equations

$$\Delta x_1 = .05x_1(1 - x_1/150,000) - \alpha x_1 x_2$$
$$\Delta x_2 = .08x_2(1 - x_2/400,000) - \alpha x_1 x_2$$

(7)

over the same state space . Here Δx_i represents the change in population x_i over a period of $\Delta t = 1$ year. We will have to supply a value for α in order to run the program. We will assume that $\alpha = 10^{-7}$ to start with. Later on we will do a sensitivity analysis on α.

Step 4 is to solve the problem by simulating the system in Eq. (7) using a computer implementation of the algorithm in Fig. 6.2. We began by simulating $N = 20$ years, starting with

$$x_1(0) = 5,000$$
$$x_2(0) = 70,000.$$

Figures 6.9 and 6.10 show the results of our first model run. Both blue whale and fin whale populations grow steadily, but in 20 years they do not get close to the equilibrium values

$$x_1 = 35,294$$
$$x_2 = 382,352$$

predicted by our analysis back in Chapter 4. Figures 6.11 and 6.12 show our simulation results when we input a value of N large enough to allow this discrete-time dynamical system to approach equilibrium.

Step 5 is to put our conclusions into plain English. It takes a long time for the whale populations to grow back—about 100 years for the fin whale, and several centuries for the more severely depleted blue whale.

We will now discuss the sensitivity of our results to the parameter α, which measures the intensity of competition between the two species. Figures 6.13 through 6.18 show the results of our simulation runs for several values of α. Of course, the equilibrium levels of both species change along with α. However, the time it takes our model to converge to equilibrium changes very little. Our general conclusion is valid whatever the extent of competition: It will take centuries for the whales to grow back.

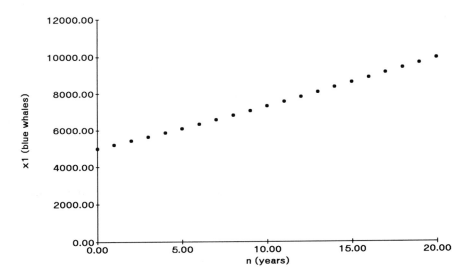

Figure 6.9 Graph of blue whales x_1 versus time n for the whale problem: case $\alpha = 10^{-7}$, $N = 20$.

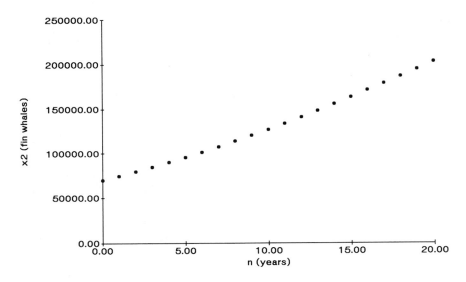

Figure 6.10 Graph of fin whales x_2 versus time n for the whale problem: case $\alpha = 10^{-7}$, $N = 20$.

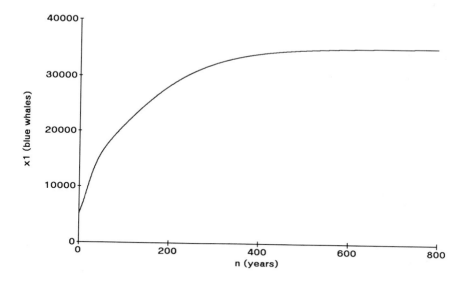

Figure 6.11 Graph of blue whales x_1 versus time n for the whale problem: case $\alpha = 10^{-7}$, $N = 800$.

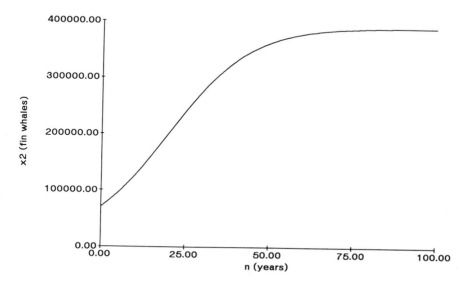

Figure 6.12 Graph of fin whales x_2 versus time n for the whale problem: case $\alpha = 10^{-7}$, $N = 100$.

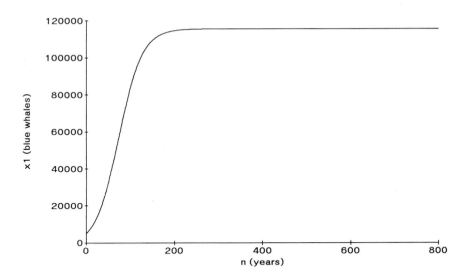

Figure 6.13 Graph of blue whales x_1 versus time n for the whale problem: case $\alpha = 3 \times 10^{-8}$, $N = 800$.

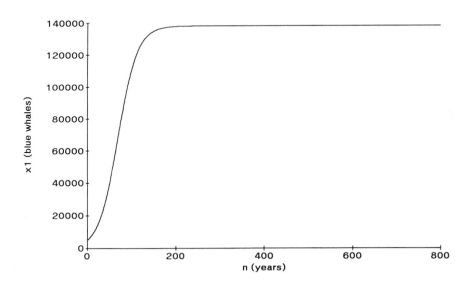

Figure 6.14 Graph of blue whales x_1 versus time n for the whale problem: case $\alpha = 10^{-8}$, $N = 800$.

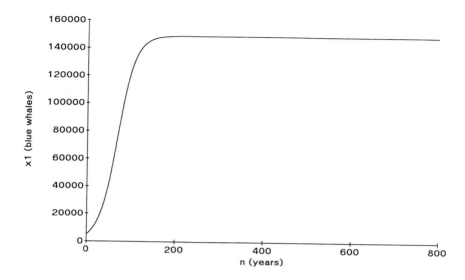

Figure 6.15 Graph of blue whales x_1 versus time n for the whale problem: case $\alpha = 10^{-9}$, $N = 800$.

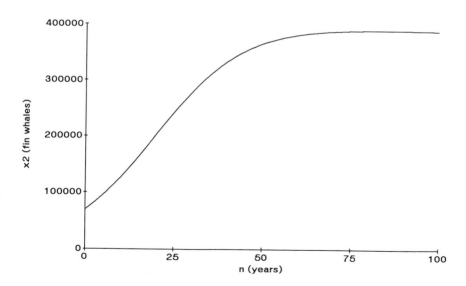

Figure 6.16 Graph of fin whales x_2 versus time n for the whale problem: case $\alpha = 3 \times 10^{-8}$, $N = 100$.

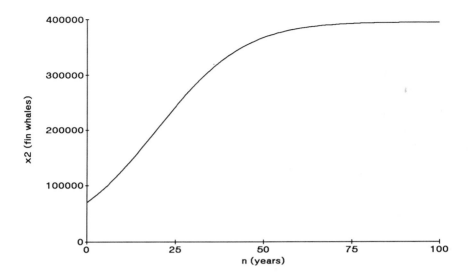

Figure 6.17 Graph of fin whales x_2 versus time n for the whale problem: case $\alpha = 10^{-8}$, $N = 100$.

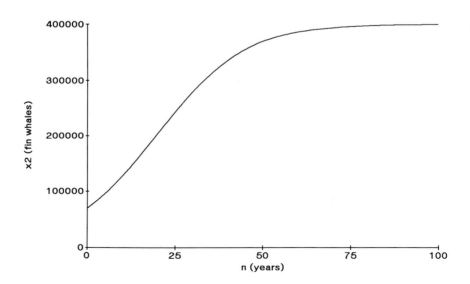

Figure 6.18 Graph of fin whales x_2 versus time n for the whale problem: case $\alpha = 10^{-9}$, $N = 100$.

6.3   The Euler Method

One of the reasons we simulate dynamic models is to obtain accurate quantitative information about system behavior. For some applications the simple simulation techniques of the previous section are too imprecise. More sophisticated numerical analysis techniques are available, however, which can be used to provide accurate solutions to initial value problems for almost any differential equation model. In this section we present the simplest generally useful method for solving systems of differential equations to any desired degree of accuracy.

Example 6.3   Reconsider the RLC circuit problem of Example 5.4 in the previous chapter. Describe the behavior of this circuit.

Our analysis in Section 5.3 was successful only in determining the local behavior of the dynamical system

$$\begin{aligned} x_1' &= x_1 - x_1^3 - x_2 \\ x_2' &= x_1 \end{aligned} \qquad (8)$$

in the neighborhood of $(0, 0)$, which is the only equilibrium of this system. The equilibrium is unstable, with nearby solution curves spiraling counterclockwise and outward. A sketch of the vector field (see Fig. 5.11) reveals little new information. There is a general counterclockwise rotation to the flow, but it is hard to tell whether solution curves spiral inward, outward, or neither, in the absence of additional information.

We will use the Euler method to simulate the dynamical system in Eq. (8). Figure 6.19 gives an algorithm for the Euler method. Consider a continuous-time dynamical system model

$$x' = F(x)$$

with $x = (x_1, \ldots, x_n)$ and $F = (f_1, \ldots, f_n)$, along with the initial condition $x(t_0) = x_0$.

Starting from this initial condition, at each iteration the Euler method produces an estimate of $x(t + h)$ based on the current estimate of $x(t)$, using the fact that

$$x(t + h) - x(t) \approx h\, F(x(t)).$$

The accuracy of the Euler method increases as the step size h becomes smaller, i.e., as the number of steps N becomes larger. For small h the error in the estimate $x(N)$ of the final value of the state variable x is roughly proportional to h. In other words, using twice as many steps (i.e. reducing h by half) produces results twice as accurate.

Algorithm: The Euler Method

Variables: $t(n)$ = time after n steps
                   $x_1(n)$ = first state variable at time $t(n)$
                   $x_1(n)$ = second state variable at time $t(n)$
                   N = number of steps
                   T = time to end simulation

Input: $t(0),\ x_1(0),\ x_2(0),\ N,\ T$

Process: Begin
                   $h \leftarrow (T - t(0))/N$
                   for $n = 0$ to $N - 1$ do
                   Begin
                       $x_1(n + 1) \leftarrow x_1(n) + hf_1(x_1(n),\ x_2(n))$
                       $x_2(n + 1) \leftarrow x_2(n) + hf_2(x_1(n),\ x_2(n))$
                       $t(n + 1) \leftarrow t(n) + h$
                       End
                   End

Output: $t(1),\ \ldots,\ t(n)$
                   $x_1(1),\ \ldots,\ x_1(N)$
                   $x_2(1),\ \ldots,\ x_2(N)$

Figure 6.19 Pseudocode for the Euler method.

Figures 6.20 and 6.21 illustrate the results obtained by applying a computer implementation of the Euler method to Eq. (8). Each graph in Figs. 6.20 and 6.21 is the result of several simulation runs. For each set of initial conditions we need to perform a sensitivity analysis on the input parameters T and N. First we enlarged T until any further enlargements produced essentially the same picture (the solution just cycled around a few more times). Then we enlarged N (i.e., decreased the step size) to check accuracy. If doubling N produced a graph that was indistinguishable from the one before, we judged that N was large enough for our purposes.

In Fig. 6.20 we started at $x_1(0) = -1$, $x_2(0) - 1.5$. The resulting solution curve spirals in toward the origin, with a counterclockwise rotation. However, before it gets too close to the origin, the solution settles into a more-or-less periodic

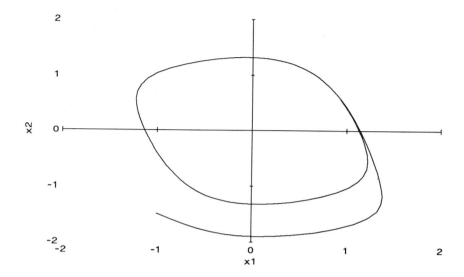

Figure 6.20 Graph of voltage x_2 versus current x_1 for the nonlinear RLC circuit problem: case $x_1(0) = -1.0$, $x_2(0) - 1.5$.

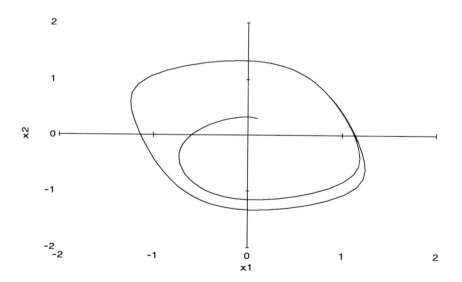

Figure 6.21 Graph of voltage x_2 versus current x_1 for the nonlinear RLC circuit problem: case $x_1(0) = 0.1$, $x_2(0) = 0.3$.

behavior, cycling around the origin. When we start nearer the origin in Fig. 6.21, the same behavior occurs, except now the solution curve spirals outward. In both cases the solution approaches the same closed loop around the origin. This closed loop is called a *limit cycle*.

Figure 6.22 shows the complete phase portrait for this dynamical system. For any initial condition except $(x_1, x_2) = (0, 0)$, the solution curve spirals counterclockwise and tends to the same limit cycle. If we begin inside the loop, the curve spirals outward; if we begin outside the loop, the curve spirals inward. The kind of behavior we see in Fig. 6.22 is a phenomenon that cannot occur in a linear dynamical system. If a solution to a linear dynamical system spirals in toward the origin, it must spiral all the way into the origin. If it spirals outward, then it spirals all the way out to infinity. This observation has modeling implications, of course. Any dynamical system not exhibiting the kind of behavior shown in Fig. 6.22 cannot be modeled adequately using linear differential equations.

Next we will perform a sensitivity analysis to determine the effect of small

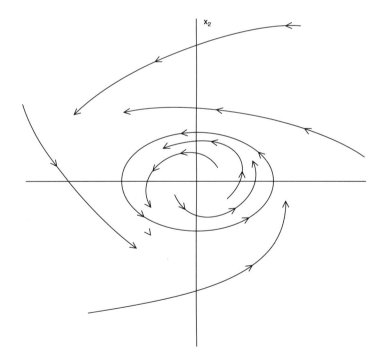

Figure 6.22 Graph of voltage x_2 versus current x_1 showing the complete phase portrait for the nonlinear RLC circuit problem of Example 6.3.

changes in our assumptions on our general conclusions. Here we will discuss the sensitivity to the capacitance C. Some additional questions of sensitivity and robustness are relegated to the exercises at the end of this chapter. In our example we assumed that $C = 1$. In the more general case we obtain the dynamical system

$$x_1' = x_1 - x_1^3 - x_2$$
$$x_2' = \frac{x_1}{C}. \tag{9}$$

For any value of $C > 0$ the vector field is essentially the same as in Fig. 5.11. The velocity vectors are vertical on the curve $x_2 = x_1 - x_1^3$ and horizontal on the x_2 axis. The only equilibrium is the origin, $(0, 0)$.

The matrix of partial derivatives is

$$A = \begin{pmatrix} 1 - 3x_1^2 & -1 \\ 1/C & 0 \end{pmatrix}.$$

Evaluate at $x_1 = 0$, $x_2 = 0$ to obtain the linear system

$$\begin{pmatrix} x_1' \\ x_2' \end{pmatrix} = \begin{pmatrix} 1 & -1 \\ 1/C & 0 \end{pmatrix} \begin{pmatrix} x_1 \\ x_2 \end{pmatrix}, \tag{10}$$

which approximates the behavior of our nonlinear system near the origin. To obtain the eigenvalues we must solve

$$\begin{vmatrix} \lambda - 1 & 1 \\ -1/C & \lambda - 0 \end{vmatrix} = 0,$$

or $\lambda^2 - \lambda + 1/C = 0$. The eigenvalues are

$$\lambda = \frac{1 \pm \sqrt{1 - 4/C}}{2}. \tag{11}$$

As long as $0 < C < 4$, the quantity under the radical is negative, so we have two complex conjugate eigenvalues with positive real parts, making the origin an unstable equilibrium.

Next we need to consider the phase portrait for the linear system. It is possible to solve the system in Eq. (10) in general by the method of eigenvalues and eigenvectors, although it would be rather messy. Fortunately in the present case it is not really necessary to determine a formula for the exact analytical solution to Eq. (10) in order to draw the phase portrait. We already know that the eigenvalues of this system are of the form $\lambda = a \pm ib$, where a is positive. As we mentioned previously (in Section 5.1, during the discussion of step 2 for Example 5.1), this implies that the coordinates of any solution curve must be linear combinations of

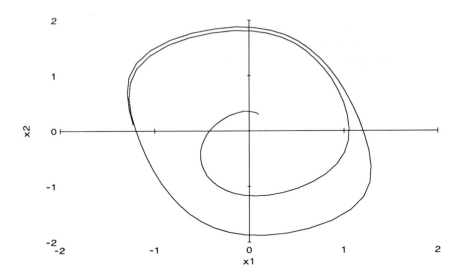

Figure 6.23 Graph of voltage x_2 versus current x_1 for the nonlinear RLC circuit problem: case $x_1(0) = 0.1$, $x_2(0) = 0.3$, $C = 0.5$.

the two terms $e^{at} \cos(bt)$ and $e^{at} \sin(bt)$. In other words, every solution curve spirals outward. A cursory examination of the vector field for Eq. (10) tells us that the spirals must rotate counterclockwise. We thus see that for any $0 < C < 4$, the phase portrait of the linear system in Eq. (10) looks much like the one in Fig. 5.10.

Our examination of the linear system in Eq. (10) shows that the behavior of the nonlinear system in the neighborhood of the origin must be essentially the same as in Fig. 6.22 for any value of C near the baseline case $C = 1$. To see what happens farther away from the origin, we need to simulate. Figures 6.23 through 6.26 show the results of simulating the dynamical system in Eq. (9) using the Euler method for several different values of C near 1. In each simulation run we started at the same initial condition as in Fig. 6.21. In each case the solution curve spirals outward and is gradually attracted to a limit cycle. The limit cycle gets smaller as C increases. Several different initial conditions were used for each value of C tested (additional simulation runs are not shown). In each case, apparently, a single limit cycle attracts every solution curve away from the origin. We conclude that the RLC circuit of Example 6.3 has the behavior shown in Fig. 6.22 regardless of the exact value of the capacitance C, assuming that C is close to 1.

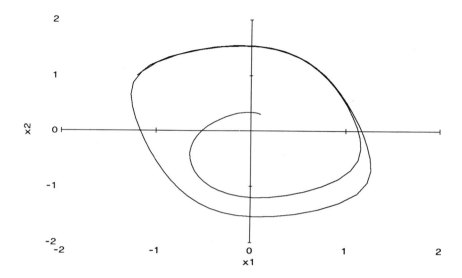

Figure 6.24 Graph of voltage x_2 versus current x_1 for the nonlinear RLC circuit problem: case $x_1(0) = 0.1$, $x_2(0) = 0.3$, $C = 0.75$.

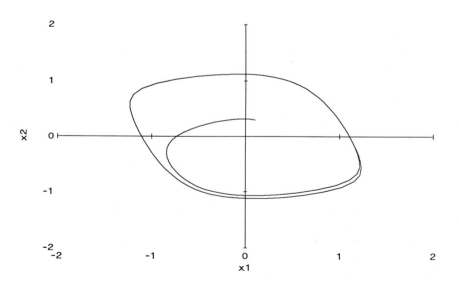

Figure 6.25 Graph of voltage x_2 versus current x_1 for the nonlinear RLC circuit problem: case $x_1(0) = 0.1$, $x_2(0) = 0.3$, $C = 1.5$.

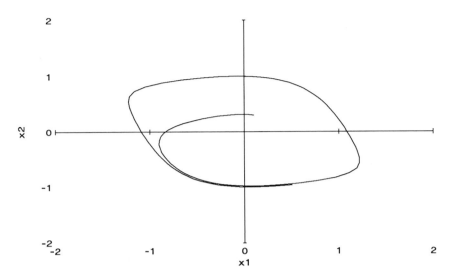

Figure 6.26 Graph of voltage x_2 versus current x_1 for the nonlinear RLC circuit problem: case $x_1(0) = 0.1$, $x_2(0) = 0.3$, $C = 2.0$.

6.4 Exercises

1. Reconsider the war problem of Example 6.1. In this problem we explore the effects of weather on combat. Bad weather and poor visibility decrease the effectiveness of direct fire weapons for both sides. The effectiveness of indirect fire weapons is relatively unaffected by the weather. We can represent the effects of bad weather in our model as follows. Let w denote the decrease in weapon effectiveness caused by bad weather conditions, and replace the dynamical system in Eq. (3) by

$$\Delta x_1 = -w\lambda(0.05)x_2 - \lambda(0.005)x_1x_2$$
$$\Delta x_2 = -w0.05x_1 - 0.005x_1x_2. \tag{12}$$

Here the parameter $0 \le w \le 1$ represents a range of weather conditions, with $w = 1$ indicating the best weather and $w = 0$ indicating the worst weather.

(a) Use a computer implementation of the algorithm in Fig. 6.2 to simulate the discrete-time dynamical system in Eq. (12) in the case $\lambda = 3$. Assume that adverse weather conditions cause a 75% decrease in weapon effectiveness for both sides ($w = 0.25$). Who wins the battle, and how long does it take? How many divisions of troops remain on the winning side?

(b) Repeat your analysis for each of the cases $w = 0.1, 0.2, 0.5, 0.75$, and 0.9, and tabulate your results. Answer the same questions as in part (a) above.

(c) Which side benefits from fighting in adverse weather conditions? If you were the blue commander, would you expect red to attack on a sunny day or a rainy day?

(d) Examine the sensitivity of your results in parts (a), (b), and (c) to the degree of weapon superiority of blue over red. Repeat the simulations in parts (a) and (b) for $\lambda = 1.5, 2.0, 4.0$, and 5.0, and tabulate your results as before. Reconsider your conclusions in part (c) above. Are they still valid?

2. Reconsider the war problem of Example 6.1. In this problem we will consider the effect of tactics on the outcome of the battle. The red commander is considering the option of holding two of his five divisions in reserve until the second or third day of combat. You can simulate each possibility as a deviation from the baseline case by running two separate simulations. First simulate the first one or two days of battle, matching two blue divisions against three red divisions. Then use the final outcome of that simulation as the initial conditions for the rest of the battle, except add two more divisions to the red side.

(a) Use a computer implementation of the algorithm in Fig. 6.2 to simulate the first phase of battle, in which two blue divisions fight three red divisions. Assume $\lambda = 2$, and tabulate the final force levels for the two cases (12 or 24 hours of battle).

(b) Use the results of part (a) to simulate the next phase of battle. Add two divisions to the final force levels for red, and continue the simulation. In each case, tell which side wins the battle, how many units remain on the winning side, and how long the battle lasts (total time for both phases of battle).

(c) The red commander may choose to commit all of his forces on the first day, or he may keep two divisions in reserve for one or two days. Which of the three strategies is better? Optimize on the basis of achieving victory at the minimum cost in terms of lost manpower.

(d) Perform a sensitivity analysis on the parameter λ, which describes the extent to which blue has weapon superiority. Repeat parts (a) and (b) for $\lambda = 1.0, 1.5, 3.0, 5.0$, and 6.0, and identify the optimal strategy for each value of λ. State your general conclusions concerning the optimal strategy for red.

3. Reconsider the war problem of Example 6.1. In this problem we will investigate the effect of tactical nuclear weapons on the battle. As a desperation

move, the blue commander considers calling for a tactical nuclear strike. It is estimated that such a strike will kill or incapacitate 70% of the red force, and 35% of the blue force as well.

(a) Use a computer implementation of the algorithm in Fig. 6.2 to simulate the discrete-time dynamical system in Eq. (3), assuming that the blue commander calls for an immediate nuclear strike. Start with initial conditions $x_1 = (0.30)5.0$, $x_2 = (0.65)2.0$, and assume $\lambda = 3$. Who wins the battle, and how long does it take? How many divisions survive on the winning side? How does blue benefit from calling for a nuclear attack in this case?

(b) Simulate the case where the blue commander waits for six hours and then calls for a nuclear attack. Simulate six hours of battle, starting with $x_1 = 5$ divisions and $x_2 = 2$ divisions. Reduce the remaining number of troops for both sides to represent the results of a nuclear strike, and then continue the simulation. Answer the same questions as in part (a).

(c) Compare the results of parts (a) and (b) to the case of conventional combat summarized in the chapter. Discuss the benefits of a tactical nuclear strike by blue. Can such a move be effective and, if so, when should the commander request the strike?

(d) Examine the sensitivity of your conclusions in part (c) to the extent λ of blue weapon superiority. Repeat the simulations of parts (a) and (b) for each of the cases $\lambda = 1.0, 1.5, 2.0, 5.0$, and 6.0, and answer the same questions as before.

4. Reconsider the space docking problem of Example 5.2. Suppose that our initial closing velocity is 50 m/sec under zero acceleration.

(a) Determine the time required for docking, assuming that the control factor $k = 0.02$. Use a computer implementation of the algorithm for discrete-time dynamical systems described in Fig. 6.2. Assume docking is complete when the closing velocity has been reduced to less than 0.1 m/sec in absolute value for all future time.

(b) Repeat the simulation of part (a) for each of the cases $k = 0.01, 0.02, 0.03, \ldots, 0.20$, and determine the time required for docking in each case. Which of these values of k results in the quickest time to dock?

(c) Repeat part (b), assuming an initial closing velocity of 25 m/sec.

(d) Repeat part (b), assuming an initial closing velocity of 100 m/sec. What conclusions can you draw about the optimal value of k for this docking procedure?

5. Reconsider the infectious disease problem introduced in Exercise 10 of Chapter 4. Use a computer implementation of the algorithm in Fig. 6.2 to simulate this discrete-time dynamical system model. Answer the questions in parts

(a) through (d) from the original exercise.

6. In Exercise 4 of Chapter 4 we introduced a simplified model of population growth in the whale problem.

    (a) Simulate this model, assuming that there are currently 5,000 blue whales and 70,000 fin whales. Use the simple simulation technique of Section 6.2, and assume $\alpha = 10^{-7}$. What happens to the two species of whales over the long term, according to this model?

    (b) Examine the sensitivity of your conclusions in part (a) to the assumption that there are currently 5,000 blue whales. Repeat the simulation of part (a), assuming that there are originally 3,000, 4,000, 6,000, or 7,000 blue whales. How sensitive are your conclusions to the exact number of blue whales in the ocean at the present time?

    (c) Examine the sensitivity of your conclusions in part (a) to the assumption that the intrinsic growth rate of the blue whale is 5% per year. Repeat the simulation of part (a), assuming that the actual rate is 3, 4, 6, or 7% per year. How sensitive are your conclusions to the actual intrinsic growth rate for the blue whales?

    (d) Examine the sensitivity of your conclusions in part (a) to the competition coefficient α. Repeat the simulation of part (a) for each of the cases $\alpha = 10^{-9}$, 10^{-8}, 10^{-6}, and 10^{-5}, and tabulate your results. How sensitive are your general conclusions to the extent of competition between the two species?

7. In Exercise 5 of Chapter 4 we introduced a more sophisticated model of population growth in the whale problem.

    (a) Simulate this model, assuming that there are currently 5,000 blue whales and 70,000 fin whales. Use the simple simulation technique of Section 6.2, and assume that $\alpha = 10^{-7}$. What happens to the two species of whales over the long term, according to this model? Do both species of whales grow back, or will one or both species become extinct? How long does this take?

    (b) Examine the sensitivity of your conclusions in part (a) to the assumption that there are currently 5,000 blue whales. Repeat the simulation of part (a), assuming that there are originally 3,000, 4,000, 6,000, 8,000, or 10,000 blue whales. How sensitive are your conclusions to the exact number of blue whales in the ocean at the present time?

    (c) Examine the sensitivity of your conclusions in part (a) to the assumption that the intrinsic growth rate of the blue whale is 5% per year. Repeat the simulation of part (a), assuming that the actual rate is 3, 4, 6, or 7% per year. How sensitive are your conclusions to the actual intrinsic growth rate for the blue whales?

(d) Examine the sensitivity of your conclusions in part (a) to the assumption that the minimum viable population level of the blue whale is 3,000 whales. Repeat the simulation of part (a), assuming that the actual level is 1,000, 2,000, 4,000, 5,000, or 6,000 whales. How sensitive are your conclusions to the actual minimum viable population level for the blue whales?

8. Reconsider Exercise 6 in Chapter 4, and assume $\alpha = 10^{-8}$. Assume that the current population levels are $B = 5,000$ and $F = 70,000$.

(a) Use a computer implementation of the simple algorithm used in Section 6.2 to determine the effect of harvesting. Assume $E = 3,000$ boat-days per year. What happens to the two species of whales over the long term, according to this model? Do both species of whales grow back, or will one or both species become extinct? How long does this take?

(b) Repeat part (a), assuming that $E = 6,000$ boat-days per year.

(c) For what range of E does the number of whales of both species approach a nonzero equilibrium?

(d) Repeat part (c) for each of the cases $\alpha = 10^{-9}$, 10^{-8}, 10^{-6}, and 10^{-5}, and tabulate your results. Discuss the sensitivity of your conclusions to the extent of interspecies competition.

9. Reconsider the whale harvesting problem of Exercise 6 in Chapter 4. In this problem we will explore the economic incentives for whalers to drive one species of whale to extinction. Assume that there are currently 5,000 blue whales and 70,000 fin whales.

(a) Simulate this model, assuming $E = 3,000$ boat-days. Use the simple simulation technique of Section 6.2, and assume that $\alpha = 10^{-7}$. Determine the long-term harvest rate in blue whale units per year (2 fin whales = 1 blue whale unit).

(b) Determine the level of effort that maximizes the long-term harvest rate in blue whale units. Simulate each of the cases $E = 500$, 1,000, 1,500, . . ., 7,500, boat-days per year. Which case results in the highest sustainable yield?

(c) Assume that whalers harvest at the rate that maximizes their long-term sustainable yield. What happens to the two species of whales over the long term, according to this model? Do both species of whales grow back, or will one or both species become extinct? How long does this take?

(d) Some economists argue that whalers will act to maximize the long-term sustainable yield for the entire industry. If so, would continued harvesting cause one or both species of whales to become extinct?

10. (Continuation of Exercise 9) Some economists argue that whalers will act in such a way as to maximize the total discounted revenue obtained by the entire whaling industry. Assume that harvesting produces revenue of $10,000 per blue whale unit and assume a discount rate of 10%. If revenue R_i is obtained in year i, the total discounted revenue is defined as

$$R_0 + \lambda R_1 + \lambda^2 R_2 + \lambda^3 R_3 + \cdots,$$

where λ represents the discount rate ($\lambda = 0.9$ for this problem).

(a) Simulate this model, assuming $E = 3,000$ boat-days. Use the simple simulation technique of Section 6.2, and assume that $\alpha = 10^{-7}$. Determine the total discounted revenue for this case.

(b) Determine the level of effort that maximizes the total discounted revenue. Simulate each of the cases $E = 500, 1,000, 1,500, \ldots, 7,500$, boat-days per year. Which case results in the highest yield?

(c) Assume that whalers harvest at the rate that maximizes their total discounted revenue. What happens to the two species of whales over the long term, according to this model? Do both species of whales grow back, or will one or both species become extinct? How long does this take?

(d) Perform a sensitivity analysis on the parameter α, which measures the extent of interspecies competition. Consider each of the cases $\alpha = 10^{-9}$, 10^{-8}, 10^{-6}, and 10^{-5}, and tabulate your results. Discuss the sensitivity of your conclusions to the extent of interspecies competition.

11. Reconsider the predator–prey model of Exercise 7 in Chapter 4.

(a) Determine by simulation the equilibrium levels for whales and krill. Use the simple simulation technique discussed in Section 6.2. Begin at several different initial conditions and run the simulation until both population levels have settled down into steady state.

(b) Suppose that after both population levels have settled down into steady state, an ecological disaster kills off 20% of the whales and 80% of the krill. Describe what happens to the two species, and how long it takes.

(c) Suppose that harvesting has depleted the whales to 5% of their equilibrium population level, while krill remain at about the same level. Describe what happens once harvesting is stopped. How long does it take for the whales to grow back? What happens to the krill population?

(d) Examine the sensitivity of your results in part (c) to the assumption that 5% of the whales remain. Simulate each of the cases where 1, 3, 7, or 10% remain, and tabulate your results. How sensitive is the time it takes for the whales to grow back to the extent to which the population is depleted?

12. Reconsider the tree problem of Example 5.1.

   (a)   Determine how long it will take for both hardwoods and softwoods to grow to 90% of their stable equilibrium levels. Assume an initial population of 1,500 tons/acre of softwood trees and 100 tons/acre of hardwoods. This is the situation in which we are trying to introduce a new type of more valuable tree into an existing ecosystem. Assume $b_i = a_i/2$, and use the simple simulation technique introduced in Section 6.2.

   (b)   Determine the point at which the biomass of hardwood trees is increasing at the fastest rate.

   (c)   Assuming that hardwoods are worth four times as much as softwoods in $/ton, determine the point at which the value of the forest stand ($/acre) is increasing at the fastest rate.

13. (Continuation of Exercise 12) Clear-cutting is a technique in which all of the trees in the forest are harvested at one time and then replanted.

   (a)   Determine the optimal harvest policy for this forest, i.e., determine the number of years we should wait before cutting and replanting. Assume that replanting involves 100 tons/acre of hardwoods and 100 tons/acre of softwoods. Base your answer on the number of $/acre per year generated.

   (b)   Determine the optimal harvest policy assuming that only hardwoods are replanted (200 tons/acre).

   (c)   Repeat part (b), but now assume that only softwoods are replanted (200 tons/acre).

   (d)   State the optimal clear-cutting policy for management of this tract of forest land. At what point would you consider selling the land rather than reforesting?

14. Reconsider the more sophisticated competing species model of Exercise 5 in Chapter 4. Assume $\alpha = 10^{-7}$.

   (a)   Use a computer implementation of the Euler method to simulate the behavior of this model, starting with the initial conditions $x_1 = 5{,}000$ blue whales and $x_2 = 70{,}000$ fin whales. Perform a sensitivity analysis on both T and N to ensure the validity of your results, as in the text. What happens to the two species of whales over the long term, according to this model? Do both species of whales grow back, or will one or both species become extinct? How long does this take?

   (b)   Repeat part (a) for a range of initial conditions for both blue and fin whales. Tabulate the results of your simulations, and answer the same questions as in part (a) for each case.

   (c)   Use the results of parts (a) and (b) in order to draw the complete phase portrait for this system.

   (d)   Indicate the region on the phase portrait where one or both species of whale are destined to become extinct.

15. Reconsider the RLC circuit problem of Example 6.3, and perform a sensitivity analysis on the parameter L, which represents inductance.

  (a) Generalize the dynamical system model in Eq. (8) to represent the case $L > 0$. How does the vector field for this model vary with L?

  (b) Determine the form of the linear system that approximates the behavior of the nonlinear RLC circuit model in the neighborhood of the origin. Calculate the eigenvalues for the linear system as a function of L. Determine the range of L over which both eigenvalues are complex with positive real part, as in the baseline case $L = 1$.

  (c) Use a computer implementation of the Euler method to simulate the behavior of the RLC circuit for each of the cases $L = 0.5, 0.75, 1.5$, and 2.0. Use the same initial condition $x_1 = 0.1$, $x_2 = 0.3$ as in Fig. 6.21. For each case, perform a sensitivity analysis on both T and N to ensure the validity of your results, as in the text.

  (d) Simulate several additional initial conditions for each value of L specified in part (c). Draw the complete phase portrait for each case. Describe how the phase portrait changes in response to changes in the inductance L.

16. Reconsider the RLC circuit problem of Example 6.3, and now consider what happens in the case of a large capacitance, $C > 4$.

  (a) Solve the linear system in Eq. (10) by the method of eigenvalues and eigenvectors in the case $C > 4$.

  (b) Draw the phase portrait for this linear system. How does the phase portrait change as a function of C?

  (c) Use a computer implementation of the Euler method to simulate the RLC circuit for each of the cases $C = 5, 6, 8$, and 10. Use the same initial condition $x_1 = 0.1$, $x_2 = 0.3$ as in Fig. 6.21. For each case, perform a sensitivity analysis on both T and N to ensure the validity of your results, as in the text.

  (d) Simulate several additional initial conditions for each value of C specified in part (c). Draw the complete phase portrait for each case. Contrast with the case $0 < C < 4$ discussed in the text. What changes occur in the phase portrait as we transition between the two cases?

17. Reconsider the RLC circuit problem of Example 6.3, and now consider the robustness of our general conclusions with respect to the assumption that the resistor in this RLC circuit has v–i characteristic $f(x) = x^3 - x$. In this problem we will assume that $f(x) = x^3 - ax$, where the parameter a may represent any positive real number. (The case $a = -4$ was the subject of Example 5.3.)

  (a) Generalize the dynamical system model in Eq. (8) to represent the general case $a > 0$. How does the vector field for this model vary with a?

(b) Determine the form of the linear system that approximates the behavior of the nonlinear RLC circuit model in the neighborhood of the origin. Calculate the eigenvalues for the linear system as a function of a.

(c) Draw the phase portrait for this linear system. How does the phase portrait change as a function of a?

(d) Use a computer implementation of the Euler method to simulate the RLC circuit for each of the cases $a = 0.5, 0.75, 1.5$, and 2.0, and draw the complete phase portrait for each case. What changes occur in the phase portrait as we change a? What do you conclude about the robustness of this model?

18. Reconsider the RLC circuit problem of Example 6.3, and now consider the robustness of our general conclusions with respect to the assumption that the resistor in this RLC circuit has $v–i$ characteristic $f(x) = x^3 - x$. In this problem we will assume that $f(x) = x^{2+b} - x$, where $b > 0$.

(a) Generalize the dynamical system model in Eq. (8) to represent the general case $b > 0$. How does the vector field for this model vary with b?

(b) Determine the form of the linear system that approximates the behavior of the nonlinear RLC circuit model in the neighborhood of the origin. Calculate the eigenvalues for the linear system as a function of b.

(c) Draw the phase portrait for this linear system. How does the phase portrait change as a function of b?

(d) Use a computer implementation of the Euler method to simulate the RLC circuit for each of the cases $b = 0.5, 0.75, 1.25$, and 1.5. Draw the complete phase portrait for each case. What changes occur in the phase portrait as we change b? What do you conclude about the robustness of this model?

19. A pendulum consists of a 100 g weight at the end of a lightweight rod 120 cm in length. The other end of the rod is fixed, but can rotate freely. The frictional forces acting on the moving pendulum are thought to be roughly proportional to its angular velocity.

(a) The pendulum is lifted manually until the rod makes a 45^0 angle with vertical. Then the pendulum is released. Determine the subsequent motion of the pendulum. Use the five-step method, and model as a continuous-time dynamical system. Simulate using the Euler method. Assume that the force due to friction is of magnitude $k\theta'$, where θ' is the angular velocity in radians per second and the coefficient of friction is $k = 0.05$ g/sec.

(b)   Use a linear approximation to determine the approximate behavior of the system near equilibrium. Assume that the magnitude of the frictional force is $k\theta'$. How does the local behavior depend on k?

(c)   Determine the period of the pendulum. How does period vary with k?

(d)   This size pendulum will be used as part of the mechanism for a grandfather clock. In order to maintain a certain amplitude of oscillation, a force is to be applied periodically. How much force should be applied, and how often, to produce an amplitude of $\pm 30^0$? How does the answer depend on the amplitude desired? [Hint: Simulate one period of the pendulum oscillation. Vary the initial angular velocity $\theta'(0)$ to obtain periodic behavior.]

20.   (Chaos) This problem illustrates the striking difference between the behavior of continuous-time and discrete-time dynamical systems that can occur even in simple models.

(a)   Show that the linear system

$$x_1' = (a - 1)x_1 - ax_1^2$$
$$x_2' = x_1 - x_2$$

has a stable equilibrium at $x_1 = x_2 = (a - 1)/a$ for any $a > 1$.

(b)   Show that the analogous discrete-time dynamical system

$$\Delta x_1 = (a - 1)x_1 - ax_1^2$$
$$\Delta x_2 = x_1 - x_2$$

also has an equilibrium at $x_1 = x_2 = (a - 1)/a$ for any $a > 1$.

(c)   Use a simulation to explore the stability of the equilibrium $x_1 = x_2 = (a - 1)/a$ and the behavior of nearby solutions for the discrete-time dynamical system. For each of the cases $a = 1.5, 2.0, 2.5, 3.0, 3.5,$ and 4.0, try several different initial conditions near the equilibrium point and report what you see. [The case $a = 4.0$ represents a simple model of chaos, the apparently random behavior of a deterministic dynamical system.]

21.   (Programming exercise) An alternative method that can be used to simulate dynamical systems is the Runge–Kutta method. Figure 6.27 gives an algorithm for the Runge–Kutta method to simulate a dynamical system in two variables,

$$\frac{dx_1}{dt} = f_1(x_1, x_2)$$
$$\frac{dx_2}{dt} = f_1(x_1, x_2).$$

Algorithm: Runge–Kutta Method

Variables: $t(n)$ = time after n steps
$x_1(n)$ = first state variable at time $t(n)$
$x_2(n)$ = second
N = number of steps
T = time to end simulation

Input: $t(0)$, $x_1(0)$, $x_2(0)$, N, T

Process: Begin
$h \leftarrow (T - t(0))/N$
for $n = 0$ to $N - 1$ do
    Begin
      $r_1 \leftarrow f_1(x_1(n), x_2(n))$
      $s_1 \leftarrow f_2(x_1(n), x_2(n))$
      $r_2 \leftarrow f_1(x_1(n) + (h/2)r_1, x_2(n) + (h/2)s_1)$
      $s_2 \leftarrow f_2(x_1(n) + (h/2)r_1, x_2(n) + (h/2)s_1)$
      $r_3 \leftarrow f_1(x_1(n) + (h/2)r_2, x_2(n) + (h/2)s_2)$
      $s_3 \leftarrow f_2(x_1(n) + (h/2)r_2, x_2(n) + (h/2)s_2)$
      $r_4 \leftarrow f_1(x_1(n) + hr_3, x_2(n) + hs_3)$
      $s_4 \leftarrow f_2(x_1(n) + hr_3, x_2(n) + hs_3)$
      $x_1(n + 1) \leftarrow x_1(n) + (h/6)(r_1 + 2r_2 + 2r_3 + r_4)$
      $x_2(n + 1) \leftarrow x_2(n) + (h/6)(s_1 + 2s_2 + 2s_3 + s_4)$
      $t(n + 1) \leftarrow t(n) + h$
    End
End

Output: $t(1), \ldots, t(N)$
$x_1(1), \ldots, x_1(N)$
$x_2(1), \ldots, x_2(N)$

Figure 6.27 Pseudocode for the Runge–Kutta method.

For a fairly small step size h, Runge–Kutta has the property that doubling the number of steps (halving h) produces results approximately 16 times more accurate.

(a)   Implement the Runge–Kutta method on a computer.

(b) Verify your computer implementation by using it to solve the linear system given by Eq. (18) in Chapter 5. Compare your results to the analytic solution in Eq. (19) for the case $c_1 = 1$, $c_2 = 0$.

(c) Verify the results obtained in Figs. 6.20 and 6.21 for the RLC circuit problem of Example 6.3.

Further Reading

Acton, F. (1970). *Numerical Methods That Work*, Harper and Row, New York.

Brams, S., Davis, M. and Straffin, P. *The Geometry of the Arms Race*, UMAP module 311.

Dahlquist, G. and Bjorck, A. *Numerical Methods*, Prentice-Hall, Englewood Cliffs, New Jersey.

Gearhart, W. and Martelli, M. *A Blood Cell Population Model, Dynamical Diseases, and Chaos*, UMAP module 709.

Press, W., Flannery, B., Teukolsky, S. and Vetterling, W. (1987). *Numerical Recipes*, Cambridge University Press, New York.

Smith, H. *Nuclear Deterrence*, UMAP module 327.

Zinnes, D., Gillespie, J. and Tahim, G. *The Richardson Arms Race Model*, UMAP module 308.

III

Probability Models

Most real-life problems contain elements of uncertainty. In some models we may introduce random elements to account for uncertainties in human behavior. In other models we may be unsure of the exact physical parameters of a system, or we may be unsure of the exact physical laws that govern its dynamics. It has even been suggested in some cases that physical parameters and physical laws are essentially random, for example, in quantum mechanics. Sometimes probabilities are introduced into a model as a matter of convenience, sometimes as a matter of necessity. In either case, it is here in the realm of probability that mathematical modeling becomes most interesting and useful.

Chapter Seven

Introduction to Probability Models

Probability is a familiar and intuitive idea. In this chapter we begin our treatment of probability models. We do not assume any prior background in formal probability theory. We will introduce the basic concepts of probability here in a natural way as they emerge in the study of real problems.

7.1 Discrete Probability Models

The most simple and intuitive probability models are those involving a discrete set of possible outcomes and no time dynamic elements. Such models are frequently encountered in the real world.

Example 7.1 An electronics manufacturer produces a variety of diodes. Quality control engineers attempt to insure that faulty diodes will be detected in the factory before they are shipped. It is estimated that 0.3% of the diodes produced will be faulty. It is possible to test each diode individually. It is also possible to place a number of diodes in series and test the entire group. If this test fails, it means that one or more of the diodes in that group are faulty. The estimated testing cost is 5 cents for a single diode, and $4 + n$ cents for a group of $n > 1$ diodes. If a group test fails, then each diode in the group must be retested individually

Variables:	n = number of diodes per test group
	C = testing cost for one group (cents)
	A = average testing cost (cents/diode)

Assumptions: If $n = 1$, then $A = 5$ cents
Otherwise $(n > 1)$, we have $C = 4 + n$
if the group test indicates that all
diodes are good, and $C = (4 + n) + 5n$
if the group test indicates a failure.

$$A = (\text{Average value of } C)/n$$

Objective: Find the value of n that minimizes A

Figure 7.1 Results of step 1 for the diode problem.

to find the bad one(s). Find the most cost-effective quality control procedure for detecting bad diodes.

We will use the five-step method. The results of step 1 are summarized in Figure 7.1. The variable n is a decision variable, and we are free to choose any $n = 1, 2, 3, \ldots$, but the variable C is the random outcome of the quality control procedure we select. We say C is a random variable. The quantity A, however, is not random. It represents the average or expected value of the random variable C/n.

Step 2 is to select the modeling approach. We will use a discrete probability model.

Consider a random variable X, which can take any of a discrete set of values

$$X \in \{x_1, x_2, x_3, \ldots\},$$

and suppose that $X = x_i$ occurs with probability p_i. We will write

$$\Pr\{X = x_i\} = p_i.$$

Of course, we must have

$$\Sigma p_i = 1.$$

Since X takes the value x_i with probability p_i, the average or expected value of X should be a weighted average of the possible values x_i, weighted according to their relative likelihoods p_i. We will write

$$EX = \Sigma x_i p_i. \tag{1}$$

The probabilities p_i represent what we will call the probability distribution of X.

Example 7.2   In a simple game of chance, two dice are rolled and the bank pays the player the number of dollars shown on the dice. How much would you pay to play this game?

Let X denote the number shown on the dice. There are $6 \times 6 = 36$ possible outcomes, and each is equally likely. There is only one way to roll a 2, so

$$\Pr\{X = 2\} = 1/36.$$

There are two ways to roll a 3 (1 and 2, or 2 and 1), so

$$\Pr\{X = 3\} = 2/36.$$

The complete probability distribution of X is illustrated in Figure 7.2. The expected value of X is

$$EX = 2(1/36) + 3(2/36) + \cdots + 12(1/36),$$

or $EX = 7$. After many repetitions of this game you would expect to win about seven dollars per roll. Therefore, it would be worthwhile to play the game if it cost no more than seven dollars to play.

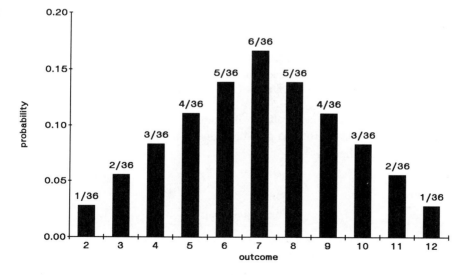

Figure 7.2 Histogram of probability versus outcome showing the distribution of the sum of two dice.

To be more specific, suppose that you play the game over and over. Let X_n denote the amount you win on the nth roll of the dice. Each X_n has the same distribution, and the different X_n are independent. The amount won on one roll does not depend on the amount won in the previous roll. There is a theorem called "the strong law of large numbers," which says that for any sequence of independent, identically distributed random variables X, X_1, X_2, X_3, ... with EX finite, we will have

$$\frac{X_1 + \cdots + X_n}{n} \to EX \tag{2}$$

as $n \to \infty$ with probability 1. In other words, if you play the game for a long time, you are virtually certain to win about \$7 per roll (Ross, S., (1985), p. 70).

The formal definition of independence is as follows. Let Y and Z denote two random variables with

$$Y \in \{y_1, y_2, y_3, \ldots\}$$

and

$$Z \in \{z_1, z_2, z_3, \ldots\}.$$

We say that Y and Z are independent if it is generally true that

$$\Pr\{Y = y_i, Z = z_j\} = \Pr\{Y = y_i\} \Pr\{Z = z_j\}. \tag{3}$$

For example, let Y and Z denote the number on the first and second dice, respectively. Then

$$\Pr\{Y = 2, Z = 1\} = \Pr\{Y = 2\} \Pr\{Z = 1\} = (1/6)(1/6) = (1/36),$$

and likewise for each possible outcome. The random variables Y and Z are independent. The number that comes up on the second die has nothing to do with what happened on the first.

Returning to the diode problem of Example 7.1, we see that the random variable C takes on one of two possible values, for any fixed $n > 1$. If all the diodes are good, then

$$C = 4 + n.$$

Otherwise

$$C = (4 + n) + 5n,$$

since we have to retest each diode. Letting p denote the probability that all the diodes are good, the remaining possibility (one or more bad diodes) must have probability $1 - p$. Then the average or expected value of C is

$$EC = (4 + n)p + [(4 + n) + 5n](1 - p). \tag{4}$$

Now for step 4. There are n diodes, and the probability that one individual diode is bad is 0.003. In other words, the probability that one individual diode is good is 0.997. Assuming independence, it follows that the probability that all n diodes in one test group are good is $p = 0.997^n$.

The expected value of the random variable C is

$$EC = (4+n)0.997^n + [(4+n) + 5n](1 - 0.997^n)$$
$$= (4+n) + 5n(1 - 0.997^n)$$
$$= 4 + 6n - 5n(0.997)^n,$$

and hence the average testing cost per diode is

$$A = 4/n + 6 - 5(0.997)^n. \tag{5}$$

The strong law of large numbers tells us that this formula represents the long-run average cost we will experience if we use test groups of size n. Now all we need is to minimize A as a function of n. We leave the details to the reader (see Exercise 1). The minimum of $A = 1.48$ cents/diode occurs at $n = 17$.

We conclude with step 5. Quality control procedures for detecting faulty diodes can be made considerably more economical by group testing methods. Individual testing costs approximately 5 cents/unit. Bad diodes occur only rarely, at a rate of 3 per 1,000. By testing groups of 17 diodes each, in series, we can reduce testing costs by a factor of three (to 1.5 cents/diode) without sacrificing quality.

Sensitivity analysis is critical in this type of problem. The implementation of a quality control procedure will depend on several factors outside the scope of our model. It may be easier to test diodes in batches of 10 or 20, or perhaps n should be a multiple of 4 or 5, depending on the details of our manufacturing process. Fortunately, the average cost A does not vary significantly between $n = 10$ and $n = 35$. Again, we leave the details to the reader. The parameter $q = 0.003$, which represents the failure rate in the manufacturing process must also be considered. For example, this value may vary with the environmental conditions inside the plant. Generalizing on our previous model, we have

$$A = 4/n + 6 - 5(1 - q)^n. \tag{6}$$

At $n = 17$ we have

$$S(A, q) = \frac{dA}{dq} \cdot \frac{q}{A} = -0.16,$$

so small variations in q are not likely to affect our cost very much.

A more general robustness analysis would consider the assumption of independence. We have assumed that there is no correlation between the times of successive failures in the manufacturing process. It may be, in fact, that bad diodes tend to be produced in batches, perhaps due to a passing anomaly in the manufacturing environment, such as a vibration or a power surge. The mathematical analysis of dependent random variable models cannot be dealt with in its entirety here. The stochastic process models introduced in the next chapter are capable of representing some kinds of dependence, while some other types of dependence admit no tractable analytic formulations. Problems in robustness are very much an active and intriguing branch of current research in probability theory. In practice, simulation results tend to indicate that expected value models based on independent random variables are quite robust. More importantly, it has been found through experience that such models provide useful, accurate approximations of real-life behavior in most cases.

7.2 Continuous Probability Models

In this section we consider probability models based on random variables that take values over a continuum. Such models are particularly convenient for representing random times. The mathematical theory required is completely analogous to the discrete case, except that now integrals replace sums.

Example 7.3   A "type I counter" is used to measure the radioactive decay in a sample of fissionable material. Decays occur at random, at an unknown rate, and the purpose of the counter is to measure the decay rate. Each radioactive decay locks the counter for a period of 3×10^{-9} seconds, during which time any decays that occur are not counted. How should the data received from the counter be adjusted to account for the lost information?

We will use the five-step method. The results of step 1 are summarized in Figure 7.3. Step 2 is to select the modeling approach. We will use a continuous probability model.

> Suppose that X is a random variable that takes values on the real line. A convenient way to describe the probability structure of X is to specify the function
> $$F(x) = \Pr\{X \le x\},$$
> called the distribution function of X. If $F(x)$ is differentiable, we call the function
> $$f(x) = F'(x)$$
> the density function of X. Then for any real numbers a and b we have
> $$\Pr\{a < X \le b\} = F(b) - F(a) = \int_a^b f(x)\,dx. \qquad (7)$$

| **Variables:** | λ = decay rate (per second) |
| | T_n = time of nth observed decay |

Assumptions:	Radioactive decays occur at random
	with rate λ. $T_{n+1} - T_n \geq 3 \times 10^{-9}$
	for all n

| **Objective:** | Find λ on the basis of a finite number |
| | of observations $T_1, \ldots, T_n$ |

Figure 7.3 Results of step 1 of the radioactive decay problem.

In other words, the area under the density curve represents probability. The mean or expected value of X is defined by

$$EX = \int_{-\infty}^{\infty} xf(x)\,dx, \tag{8}$$

which is directly analogous to the discrete case, as can be seen by considering the Riemann sum for the integral. (See Exercise 13 for details.) It is also worthwhile pointing out that this notation and terminology were originally adapted from a problem in physics, namely the center of mass problem. If a wire or rigid rod is laid out along the x axis, and $f(x)$ represents the density (gms/cm) at point x, then the integral of $f(x)$ represents mass, and the integral of $xf(x)$ represents the center of mass (assuming, as we do in probability, that the total mass is equal to 1).

The special case of random arrivals occurs frequently in applications. Suppose that arrivals (e.g., customers, phone calls, radioactive decays) occur at random with rate λ, and let X denote the random time between two successive arrivals. It is common to assume that X has the distribution function

$$F(t) = 1 - e^{-\lambda t}, \tag{9}$$

so that the density function of X is

$$f(t) = \lambda e^{-\lambda t}. \tag{10}$$

This distribution is called the *negative exponential distribution* with rate parameter λ.

One very important property of the negative exponential distribution is its "lack of memory." For any $t > 0$ and $s > 0$, we have

$$
\begin{aligned}
\Pr\{X > s + t \mid X > s\} &= \frac{\Pr\{X > s + t\}}{\Pr\{X > s\}} \\
&= \frac{e^{-\lambda(s+t)}}{e^{-\lambda s}} = e^{-\lambda t} \\
&= \Pr\{X > t\}.
\end{aligned}
\tag{11}
$$

In other words, the fact that we have already waited s units of time for the next arrival does not affect the (conditional) distribution of the time until the next arrival. The negative exponential distribution "forgets" that we have already waited this long. The probability in Eq. (11) is called a conditional probability. Formally, the probability of event A occurring, given the event B occurs, is

$$
\Pr\{A \mid B\} = \frac{\Pr\{A \text{ and } B\}}{\Pr\{B\}}.
\tag{12}
$$

In other words, $\Pr\{A \mid B\}$ is the relative likelihood of A among all possible events once B has occurred.

Now we proceed to step 3, the model formulation. We are assuming that radioactive decays occur at random at an unknown rate λ. We will model this process by assuming that the times between successive radioactive decays are independent and identically distributed with a negative exponential distribution with rate parameter λ. Let

$$
X_n = T_n - T_{n-1}
$$

denote the times between successive observations of a radioactive decay. Of course, X_n does not have the same distribution as the time between successive decays, because of the lock time. In fact $X_n \geq 3 \times 10^{-9}$ with probability 1, which is certainly not true for the negative exponential distribution.

The random time X_n consists of two parts. First we must wait $a = 3 \times 10^{-9}$ seconds while the counter is locked, and then we must wait an additional Y_n seconds until the next decay. Now Y_n is not simply the time between two decays, because it begins at the end of the lock time, not at a decay time. However, the memoryless property of the negative exponential distribution guarantees that Y_n is still negative exponential with rate parameter λ.

Step 4 is to solve the model. Since $X_n = a + Y_n$, we have $E X_n = a + E Y_n$, where

$$
E Y_n = \int_0^\infty t \lambda e^{-\lambda t} \, dt.
$$

Integrate by parts to find $E Y_n = 1/\lambda$. Thus, $E X_n = a + 1/\lambda$. The strong law of large numbers says that

$$
\lim_{n \to \infty} \frac{X_1 + \cdots + X_n}{n} = a + \frac{1}{\lambda}
$$

with probability 1. In other words, $(T_n/n) \to a + 1/\lambda$. For large n it will be approximately true that

$$\frac{T_n}{n} = a + \frac{1}{\lambda}. \tag{13}$$

Solving for λ, we obtain

$$\lambda = \frac{n}{T_n - na}. \tag{14}$$

Finally, step 5. We have obtained a formula for decay rate which corrects for the decays missed while the counter is locked. All that is required is to record the length of observation and the number of decays recorded. The distribution of those decays in the observation interval is not required to determine λ.

Sensitivity analysis should consider the lock time a, which must be determined empirically. The accuracy to which we can determine a will affect the accuracy of λ. From Eq. (14) we calculate that

$$\frac{d\lambda}{da} = \lambda^2.$$

The sensitivity of λ to a is then

$$S(\lambda, a) = \lambda^2(a/\lambda) = \lambda a.$$

This is also the expected number of decays during the lock time. We can therefore get a better estimate of λ (in relative terms) for a less intensely radioactive source. One simple way to achieve this is to use fewer grams of radioactive material in our sample. Another important source of potential error comes from the assumption that

$$(X_1 + \cdots + X_n)/n = a + 1/\lambda.$$

Of course this is not exactly true. Random fluctuations will cause the empirical rate to vary from the mean, although we do have convergence as $n \to \infty$. The study of such random fluctuations is the subject of the next section.

Finally there is the matter of robustness. We have made an assumption about the decay process which appears to be very special. We have assumed that times between decays are independent and that they have a particular distribution (negative exponential with rate parameter λ). Such an arrival process is called a *Poisson process*. The Poisson process is commonly used to represent random arrivals. Its use can be justified in part by the fact that many real-world arrival processes have interarrival times that are at least approximately negative exponential. This can be verified by collecting data on arrival times. But this does not answer the question of *why* the negative exponential distribution occurs.

It turns out that there is a mathematical reason for expecting arrival processes to look Poisson. Consider a large number of arrival processes that are independent of one another. We make no assumption about the interarrival time distribution of an arrival process, only that the interarrival times are independent and identically distributed. There is a theorem that states, under fairly general conditions, that the arrival process obtained by merging all of these independent processes has to look Poisson. (The merged process tends to Poisson as the number of merged processes tends to infinity.) This is why the Poisson process, based on the negative exponential distribution, is such a robust model (See Feller, W., (1971), p. 370).

7.3   Introduction to Statistics

In any modeling situation it is desirable to get quantitative measures of performance. For probability models an additional complication is involved in deriving such parameters of system behavior. We must have a way to deal with the random fluctuations in system behavior that are characteristic of probability models. Statistics is the study of measurement in the presence of random fluctuations. The appropriate use of statistical methods must be a part of the analysis of any probability model.

Example 7.4   An emergency 911 service in a local community received an average of 171 calls per month for house fires over the past year. On the basis of this data the rate of house fire emergencies was estimated at 171 per month. The next month only 153 calls were received. Does this indicate an actual reduction in the rate of house fires, or is it simply a random fluctuation?

We will use the five-step method. The results of step 1 are summarized in Figure 7.4. We are assuming negative exponential interarrival times for the emergency calls. Step 2 is to determine the modeling approach. We will model this as a statistical inference problem.

Suppose that X, X_1, X_2, X_3, ..., are independent random variables, all with the same distribution. Recall that if X is discrete, the average or expected value is

$$EX = \Sigma \, x_k \Pr\{X = x_k\},$$

and if X is continuous with density $f(x)$, then

$$EX = \int xf(x)\,dx.$$

Another distributional parameter, called the *variance*, measures the extent to which X tends to deviate from the mean EX. In general we define

$$VX = E(X - EX)^2. \tag{15}$$

Variables: λ = Rate of house fire reports (per month)
X_n = time between $(n-1)$st and nth fire (months)

Assumptions: House fires occur at random with rate λ,
i.e., $X_1,\ X_2, \ldots$ are independent and each X_n has
a negative exponential distribution with rate
parameter λ.

Objective: Determine the probability that as few as 153 calls
would be received in one month, given $\lambda = 171$.

Figure 7.4 Results of step 1 of the house fire problem.

If X is discrete we have

$$VX = \Sigma(x_k - EX)^2 \Pr\{X = x_k\}, \tag{16}$$

and if X is continuous with density $f(x)$, we have

$$VX = \int (x - EX)^2 f(x)\, dx. \tag{17}$$

There is a result called the *central limit theorem*, which states that as $n \to \infty$ the distribution of the sum $X_1 + \cdots + X_n$ gets closer and closer to a certain type of distribution called a *normal distribution*. Specifically, if we let $\mu = EX$ and $\sigma^2 = VX$, then for all t real we have

$$\lim_{n\to\infty} \Pr\left\{ \frac{X_1 + \cdots + X_n - n\mu}{\sigma\sqrt{n}} \le t \right\} \to \Phi(t) \tag{18}$$

where $\Phi(t)$ is a special distribution function called the *standard normal distribution*. The density function for the standard normal is defined for all x by

$$g(x) = \frac{1}{\sqrt{2\pi}} e^{-x^2/2}, \tag{19}$$

so for all t we have

$$\Phi(t) = \int_{-\infty}^{t} \frac{1}{\sqrt{2\pi}} e^{-x^2/2}\, dx. \tag{20}$$

Figure 7.5 shows a graph of the standard normal density. Numerical integration shows that the area between $-1 \le x \le 1$ is approximately 0.68, and the area between $-2 \leq x \leq 2$ is approximately 0.95. Thus, for all n sufficiently large we will have

$$-1 \leq \frac{X_1 + \cdots + X_n - n\mu}{\sigma\sqrt{n}} \leq 1$$

about 68% of the time, and

$$-2 \leq \frac{X_1 + \cdots + X_n - n\mu}{\sigma\sqrt{n}} \leq 2 \qquad (21)$$

about 95% of the time. In other words, we are 68% sure that

$$n\mu - \sigma\sqrt{n} \leq X_1 + \cdots + X_n \leq n\mu + \sigma\sqrt{n},$$

and 95% sure that

$$n\mu - 2\sigma\sqrt{n} \leq X_1 + \cdots + X_n \leq n\mu + 2\sigma\sqrt{n}. \qquad (22)$$

It is common in practice to accept the 95% interval from Eq. (22) as the range of normal variation in a random sample. In cases where the sum $X_1 + \cdots + X_n$ does not lie in the interval from Eq. (22), we say that the deviation is *statistically significant* at the 95% level.

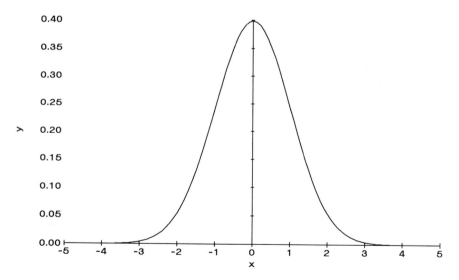

Figure 7.5 Graph of the standard normal density function.

$$g(x) = \frac{1}{\sqrt{2\pi}} e^{-x^2/2}$$

Now we move on to step 3, the model formulation. We are assuming that the times between calls, X_n, are negative exponentially distributed, with density function

$$f(x) = \lambda e^{-\lambda x}$$

on $x \geq 0$. We have previously calculated that

$$\mu = E X_n = 1/\lambda.$$

The variance

$$\sigma^2 = V X_n$$

is given by

$$\sigma^2 = \int_0^\infty (x - 1/\lambda)^2 \lambda e^{-\lambda x}\, dx,$$

and the central limit theorem gives probability estimates of the extent to which

$$(X_1 + \cdots + X_n)$$

can vary from its mean, n/λ. In particular we know that Eq. (22) holds with probability 0.95.

On to step 4. We calculate that

$$\sigma^2 = 1/\lambda^2,$$

using integration by parts. Substituting $\mu = 1/\lambda$ and $\sigma = 1/\lambda$ into Eq. (22), we find that the relation

$$n/\lambda - 2\sqrt{n}/\lambda \leq X_1 + \cdots + X_n \leq n/\lambda + 2\sqrt{n}/\lambda \tag{23}$$

must hold with probability 0.95. Substituting $\lambda = 171$ and $n = 153$ into Eq. (23), we are 95% sure that

$$153/171 - 2\sqrt{153}/171 \leq X_1 + \cdots + X_{153} \leq 153/171 + 2\sqrt{153}/171,$$

or in other words,

$$0.75 \leq X_1 + \cdots + X_{153} \leq 1.04.$$

Therefore, our observation that

$$X_1 + \cdots + X_{153} \approx 1$$

is within the range of normal variation.

Finally, step 5. There is insufficient evidence to conclude that the rate of house fire emergency calls has declined. The variation in the observed number of calls may well be the result of normal random variation. Of course, if the observed number of calls per month continues to be this low, then we would reassess the situation.

A few items should be included in our sensitivity analysis. First of all, we have concluded that 153 calls in one month is within the range of normal variation. More generally, suppose that n calls are received in one month. Substituting $\lambda = 171$ into Eq. (23), we conclude that

$$n/171 - 2\sqrt{n}/171 \leq X_1 + \cdots + X_n \leq n/171 + 2\sqrt{n}/171 \qquad (24)$$

with probability 0.95. Since the interval

$$n/171 \pm 2\sqrt{n}/171$$

contains 1 for any value of $n \in [147, 198]$, we would conclude more generally that 95% of the time there will be between 147 and 198 calls in a month. In other words, the range of normal variation for this community is 147 to 198 emergency calls per month.

Now let us consider the sensitivity of our conclusions to the assumption that the actual expected number of emergency calls is 171 per month. Less specifically, assume that there are an average of λ emergency calls per month. We have observed $n = 153$ calls in a one-month period. Substituting into Eq. (23), we conclude that

$$153/\lambda - 2\sqrt{153}/\lambda \leq X_1 + \cdots + X_{153} \leq 153/\lambda + 2\sqrt{153}/\lambda \qquad (25)$$

with probability 0.95. Since the interval

$$153/\lambda \pm 2\sqrt{153}/\lambda$$

contains 1 for any value of λ between 128 and 178, we conclude that a month with 153 emergency calls is within the range of normal variation for any community in which the average number of emergency calls is between 128 and 178 per month.

There is a final matter of robustness that requires comment. We have assumed that the times between calls, X_n, are negative exponential. However, the central limit theorem remains true for any distribution as long as μ and σ are finite. Hence, our conclusion is really not sensitive to the assumption of a negative exponential distribution. It only requires that σ is not too much smaller than μ (for a negative exponential, $\mu = \sigma$). As we remarked at the end of Section 7.2, there is a good reason to expect this to be the case. Of course, we can always check by estimating μ and σ from the data.

7.4 Exercises

1. Consider the diode problem of Example 7.1. Let

$$A(x) = 4/x + 6 - 5(0.997)^x$$

denote the average cost function.
(a) Show that on the interval $x > 0$, $A(x)$ has a unique minimum value at the point where $A'(x) = 0$.
(b) Use a numerical method to estimate the minimum to within 0.1.
(c) Find the minimum value of $A(x)$ over the set $x = 1, 2, 3, \ldots$.
(d) Find the maximum of $A(x)$ over the set $10 \le x \le 35$.

2. Reconsider the diode problem of Example 7.1. In this exercise we will investigate the problem of estimating the failure rate q. In Example 7.1 we assumed that $q = 0.003$, in other words that 3 out of 1,000 diodes are flawed.
(a) Suppose that we test a batch of 1,000 diodes and we find that 3 are flawed. On this basis we estimate that $q = 3/1,000$. How accurate is this estimate? Use the five-step method, and model as a statistical inference problem. [Suggestion: Define a random variable X_n to represent the status of the nth diode. Let $X_n = 0$ if the nth diode is good, or $X_n = 1$ if the nth diode has a flaw. Then the sum

$$\frac{X_1 + \cdots + X_n}{n}$$

represents the fraction of bad diodes, and the central limit theorem can be used to estimate the likely variation of this sum from the true mean q.]
(b) Repeat part (a), assuming that 30 bad diodes were found in a batch of 10,000.
(c) How many diodes need to be tested in order to be 95% sure that we have determined the failure rate to within 10% of its true value? Assume that the true value is close to our original estimate of $q = 0.003$.
(d) How many diodes need to be tested in order to be 95% sure that we have determined the failure rate to within 1% of its true value? Assume that the true value is close to our original estimate of $q = 0.003$.

3. Consider the radioactive decay problem of Example 7.3, and suppose that our counter registers an average of 10^7 decays/sec over a period of 30 seconds.
(a) Use Eq. (14) to estimate the actual decay rate λ.
(b) Generalize Eq. (13), using the central limit theorem. Calculate the range of normal variation (at the 95% level) for the observed decay rate T_n/n for an arbitrary value of the true decay rate λ.

(c)   Determine the range of λ for which the observed decay rate $T_n/n = 10^7$ is within the range of normal variation at the 95% level. How accurate is our estimate in part (a)?

(d)   How long would we have to sample this radioactive material to be 95% sure we have determined the true decay rate λ to six significant digits (i.e., to within $0.5 \times 10^{-6} \lambda$)?

4.   It can be shown that the area beneath the graph of the standard normal density function (see Eq. (19)) between $-3 \le x \le 3$ is approximately 0.997. In other words, for large n we are about 99.7% sure that

$$n\mu - 3\sigma\sqrt{n} \le X_1 + \cdots + X_n \le n\mu + 3\sigma\sqrt{n}.$$

Use this fact to repeat the calculations in Exercise 3, but use the 99.7% confidence level. Comment on the sensitivity of your answer in part (d) to the confidence level.

5.   Reconsider the house fire problem of Example 7.4. In this exercise we will investigate the problem of estimating the rate λ at which emergency calls occur.

(a)   Suppose that 2,050 emergency calls are received in a one-year period. Estimate the rate λ of house fires per month.

(b)   Assuming that the true value of λ is 171 calls per month, calculate the range of normal variation for the number of emergency calls received in one year.

(c)   Calculate the range of λ for which 2,050 calls in one year is within the range of normal variation. How accurate is our estimate of the true rate λ at which house fires occur?

(d)   How many years of data would be required to obtain an estimate of λ accurate to the nearest integer (an error of ± 0.5)?

6.   Reconsider the house fire problem of Example 7.4. The underlying random process is called a Poisson process because it can be shown that the number of arrivals (calls) N_t during a time interval of length t has a Poisson distribution. Specifically

$$\Pr\{N_t = n\} = e^{-\lambda t}(\lambda t)^n/n!$$

for all $n = 0, 1, 2, \ldots$.

(a)   Show that

$$E N_t = \lambda t$$

and

$$V N_t = \lambda t.$$

(b)   Use the Poisson distribution to calculate the probability that the number of calls received in a given month deviates from the mean of 171 by as much as 18 calls.

(c)   Generalize the calculation of part (b) to determine the exact range of normal variation (at the 95% level) for the number of calls in a one-month period.

(d)   Compare the exact method used in part (c) with the approximate calculation of the range of normal variation that is included in the discussion of sensitivity analysis for Example 7.4 in the text. Which method would be more appropriate for determining the range of normal variation in the number of calls received in a single day? In a year?

7.   The Michigan state lottery runs a game in which you pay $1 to buy a ticket containing a three-digit number of your choice. If your number is drawn at the end of the day, you win $500.

(a)   Suppose you were to buy one ticket per week for a year. What are your chances of coming out a winner for the year? [Hint: It is easy to compute the probability of coming out a loser!]

(b)   Can you improve your chances of coming out a winner this year by purchasing more than one ticket per week? Calculate the probability of coming out a winner if you buy n tickets a week, for $n = 1, 2, 3, \ldots, 9$.

(c)   Suppose that the state lottery sells 1,000,000 tickets this week. What is the range of likely variation in the amount of money the state will make this week? How likely is it that the state will lose money this week? Use the central limit theorem.

(d)   What is wrong with using the central limit theorem to answer the question in part (a) or (b)?

8.   (Murphy's Law–part I) You are staying at a downtown hotel. In front of the hotel there is a taxicab stand. Taxis arrive at random, at a rate of about one every five minutes.

(a)   How long do you expect to wait for a taxicab, assuming that there is none at the hotel when you exit?

(b)   The time until the arrival of the next taxicab is called a *forward recurrence time*. The time since the most recent arrival is called a *backward recurrence time*. For a Poisson process it can be shown that backward and forward recurrence times have the same distribution. (The probabilistic behavior of the process is the same if we let time run in reverse.) Using this fact, how long on average has it been since the last taxi arrived at the time you exit the hotel?

(c)   On average, the length of time between taxicab arrivals is five minutes. How long on average is the length of time between the arrival of the cab you just missed and the one you have to wait for?

9. (Murphy's Law–part II) You are at the supermarket checkout stand. After waiting for what seems like an unusually long time to check out, you decide to conduct a scientific experiment. One by one, you measure the length of time each customer has to wait. You continue until you find one who has to wait longer than you did.

   (a) Let X denote the time you had to wait, and let X_n denote the time the nth customer had to wait. Let N denote the number of the first customer n for which $X_n \geq X$. To be fair, assume that $X, X_1, X_2, \ldots$ are identically distributed. Explain why the probability that $N \geq n$ (i.e., that out of the group consisting of you and the first $n - 1$ customers, you waited the longest) must equal $1/n$.

   (b) Calculate the probability distribution of the random variable N.

   (c) Calculate the expected value EN, which represents the number of customers you must observe, on average, until you find one who waited longer than you did.

10. (Murphy's Law–part III) A doctor of internal medicine with a busy practice expects to be called into the hospital to respond to a serious heart attack on average about once every two weeks. Assume that heart attacks in this physician's patient population occur at random with this rate. One such emergency call is a challenge. Two such calls in a single day is a disaster.

    (a) How many heart attacks should the physician expect to respond to in a single year?

    (b) Explain why the probability that n heart attacks during the year all occur on different days is

    $$\frac{365}{365} \cdot \frac{364}{365} \cdot \frac{363}{365} \cdots \frac{365 - n + 1}{365}.$$

    (c) What is the probability that the doctor has to respond to two or more heart attacks on the same day some time this year?

11. A squadron of 16 bombers needs to penetrate air defenses to reach its target. They can either fly low and expose themselves to the air defense guns, or fly high and expose themselves to surface-to-air missiles. In either case, the air defense firing sequence proceeds in three stages. First they must detect the target, then they must acquire the target (lock on target), and finally they must hit the target. Each of these stages may or may not succeed. The probabilities are as follows:

AD Type	P_{detect}	$P_{acquire}$	P_{hit}
Low	0.90	0.80	0.05
High	0.75	0.95	0.70

The guns can fire 20 shells per minute, and the missile installation can fire 3 per minute. The proposed flight path will expose the planes for 1 minute if they fly low, and 5 minutes if they fly high.

(a)   Determine the optimal flight path—low or high. The objective is to maximize the number of bombers that survive to strike the target.

(b)   Each individual bomber has a 70% chance to destroy the target. Use the results of part (a) to determine the chances of success (target destroyed) for this mission.

(c)   Determine the minimum number of bombers necessary to guarantee a 95% chance of mission success.

(d)   Perform a sensitivity analysis with respect to the probability $p = 0.7$ that an individual bomber can destroy the target. Consider the number of bombers that must be sent to guarantee a 95% chance of mission success.

(e)   Bad weather reduces both P_{detect} and p, the probability that a bomber can destroy the target. If all of these probabilities are reduced in the same proportion, which side gains an advantage in bad weather?

12.   A scanning radio communications sensor attempts to detect radio emissions and pinpoint their locations. The sensor scans 4,096 frequency bands. It takes the sensor 0.1 seconds to detect a signal. If no signal is detected, it moves to the next frequency. If a signal is detected, it takes an additional 5 seconds to get a location fix. There is no signal except on about 100 of the frequency bands, but the sensor does not know which ones are being used, so it must scan them all. On the busy frequencies, the percent utilization (i.e., the fraction of time that the signal is on) varies from 30% to 70%. An additional complication is that emissions on the same frequency can come from several different sources, so the sensor must continue to scan all frequencies even after a source is located.

(a)   Determine the approximate detection rate for this system. Assume that all frequency bands are scanned sequentially.

(b)   Suppose the sensor has the capability of remembering 25 high-priority frequency bands, which are scanned 10 times as often as the others. Assuming that the sensor is eventually able to identify 25 busy frequencies and gives them high priority, how does the detection rate change?

(c)   Suppose that in order to obtain useful information we must be able to detect emissions on a particular frequency at a minimum rate of once every three minutes. Determine the optimal number of high-priority channels.

(d)   Perform a sensitivity analysis on the answer to part (c) with respect to the average utilization rate for busy frequencies.

13. In this problem we will explore the parallel between discrete and continuous random variables. Suppose that X is a continuous random variable with distribution function $F(x)$ and density function $f(x) = F'(x)$. For each n we define a discrete random variable X_n with approximately the same distribution as X. Partition the real line into intervals of length $\Delta x = n^{-1}$, and let I_i denote the ith interval in the partition. Then for each i, select a point x_i in the ith interval, and define

$$p_i = \Pr\{X_n = x_i\} = f(x_i)\Delta x.$$

(a) Explain why we can always choose the points x_i so that we will have

$$p_i = \Pr\{X \in I_i\}$$

for all i. This ensures that $\sum p_i = 1$.

(b) Derive a formula that represents the probability that $a < X_n \leq b$ in terms of the density function f for any two real numbers a and b.

(c) Derive a formula that represents the mean EX_n in terms of the density function f.

(d) Use the results of part (b) to show that as $n \to \infty$ (or, equivalently, $\Delta x \to 0$), we have

$$\Pr\{a < X_n \leq b\} \to \Pr\{a < X \leq b\}$$

for any two real numbers a and b. We say that X_n *converges in distribution* to X.

(e) Use the results of part (c) to show that as $n \to \infty$ (or, equivalently, $\Delta x \to 0$), we have

$$EX_n \to EX.$$

We say that X_n *converges in mean* to X.

Further Reading

Barnier, W. *Expected Loss in Keno*, UMAP module 574.

Berresford, G. *Random Walks and Fluctuations*, UMAP module 538.

Billingsley, P. (1979). *Probability and Measure*, Wiley, New York.

Carlson, R. *Conditional Probability and Ambiguous Information*, UMAP module 391.

Feller, W. (1971). *An Introduction to Probability Theory and Its Applications*, Vol. 2, 2nd ed., Wiley, New York.

Moore, P. and McCabe, G. (1989). *Introduction to the Practice of Statistics* W.H. Freeman, New York.

Ross, S. (1985). *Introduction to Probability Models*, 3rd ed., Academic Press, New York.

Watkins, S. *Expected Value at Jai Alai and Pari-Mutuel Gambling*, UMAP module 631.

Wilde, C. *The Poisson Random Process*, UMAP module 340.

Chapter Eight

Stochastic Models

The deterministic dynamic models of part II of this book do not allow for the explicit representation of uncertainty. When random effects are taken into account, the resulting model is called a *stochastic model*. Several kinds of general stochastic models are in wide use today. In this chapter we will introduce the most important and commonly used stochastic models.

8.1 Markov Chains

A Markov chain is a discrete-time stochastic model. It is a generalization of the discrete-time dynamical system model introduced in Section 4.3. Although the model is simple, the number and diversity of applications are surprisingly large. In this section we will introduce the general Markov chain model. We will also introduce a concept of steady state appropriate for a stochastic model.

Example 8.1 A pet store sells a limited number of 20-gallon aquariums. At the end of each week the store manager takes inventory and places orders. Store policy is to order 3 new 20-gallon aquariums at the end of the week if all of the current inventory has been sold. If even 1 of the 20-gallon aquarium remains in stock, no new units are ordered. This policy is based on the observation that the store only sells an average of 1 of the 20-gallon aquarium per week. Is this

| **Variables:** | $S_n = $ supply of aquariums at the beginning of week n |
| | $D_n = $ demand for aquariums during week n |

Assumptions: If $D_{n-1} < S_{n-1}$, then $S_n = S_{n-1} - D_{n-1}$
If $D_{n-1} \geq S_{n-1}$, then $S_n = 3$
$\Pr\{D_n = k\} = e^{-1}/k!$

Objective: Calculate $\Pr\{D_n > S_n\}$

Figure 8.1 Results of step 1 of the inventory problem.

policy adequate to guard against potential lost sales of 20-gallon aquariums due to a customer requesting one when they are out of stock?

We will use the five-step method. Step 1 is to ask a question. The store begins each sales week with an inventory of between 1 and 3 of the 20-gallon aquariums. The number of sales in one week depends on both the supply and the demand. The demand averages one per week but is subject to random fluctuations. It is possible that on some weeks demand will exceed supply, even if we start the week with the maximum inventory of three units. We would like to calculate the probability that demand exceeds supply. In order to get a specific answer, we need to make an assumption about the probabilistic nature of the demand. It seems reasonable to assume that potential buyers arrive at random at a rate of one per week. Hence, the number of potential buyers in one week will have a Poisson distribution with mean one. (The Poisson distribution was introduced in Exercise 6 of Chapter 7.) Figure 8.1 summarizes the results of step 1.

Step 2 is to select the modeling approach. We will use a Markov chain model.

A *Markov chain* can best be described as a sequence of random jumps. For the purposes of this book we will assume that these jumps can only involve a finite discrete set of locations or states. Suppose that the random variables X_n take values in a finite discrete set. There is no harm in assuming that

$$X_n \in \{1, 2, 3, \ldots, m\}.$$

We say that the sequence $\{X_n\}$ is a Markov chain provided that the probability that $X_{n+1} = j$ depends only on X_n. If we define

$$p_{ij} = \Pr\{X_{n+1} = j \mid X_n = i\}, \tag{1}$$

then the entire future history of the process $\{X_n\}$ is determined by the p_{ij} and the probability distribution of the initial X_0. Of course, when we say "determined," we mean that the probabilities $\Pr\{X_n = i\}$ are determined. The actual value of X_n depends on random factors.

Example 8.2   Describe the behavior of the following Markov chain. The state variable

$$X_n \in \{1, 2, 3\}.$$

If $X_n = 1$, then $X_{n+1} = 1$, 2, or 3 with equal probability. If $X_n = 2$, then $X_{n+1} = 1$ with probability 0.7, and $X_{n+1} = 2$ with probability 0.3. If $X_n = 3$, then $X_{n+1} = 1$ with probability 1.

The state transition probabilities p_{ij} are given by

$$p_{11} = \frac{1}{3}$$

$$p_{12} = \frac{1}{3}$$

$$p_{13} = \frac{1}{3}$$

$$p_{21} = 0.7$$

$$p_{22} = 0.3$$

$$p_{31} = 1,$$

and the rest are zero. It is customary to write the p_{ij} in matrix form:

$$P = (p_{ij}) = \begin{pmatrix} p_{11} & \cdots & p_{1m} \\ \vdots & & \vdots \\ p_{m1} & \cdots & p_{mm} \end{pmatrix}. \tag{2}$$

Here

$$P = \begin{pmatrix} 1/3 & 1/3 & 1/3 \\ 0.7 & 0.3 & 0 \\ 1 & 0 & 0 \end{pmatrix}.$$

Another convenient method is called the *state transition diagram* (see Figure 8.2). This makes it easy to visualize the Markov chain as a sequence of random jumps. Suppose $X_0 = 1$. Then $X_1 = 1$, 2, or 3 with probability $1/3$ each. The probability that $X_2 = 1$ is obtained by calculating the probability associated with each individual sequence of jumps that transitions from state 1 to state 1 in two steps. Thus,

$$\Pr\{X_2 = 1\} = \left(\frac{1}{3}\right)\left(\frac{1}{3}\right) + \left(\frac{1}{3}\right)(0.7) + \left(\frac{1}{3}\right)(1) = 0.67\overline{7}.$$

Similarly,

$$\Pr\{X_2 = 2\} = \left(\frac{1}{3}\right)\left(\frac{1}{3}\right) + \left(\frac{1}{3}\right)(0.3) = 0.21\overline{1},$$

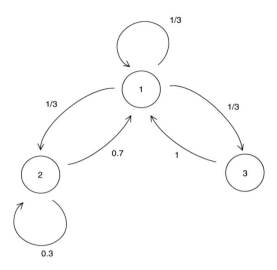

Figure 8.2 State transition diagram for Example 8.2.

and

$$\Pr\{X_2 = 3\} = \left(\frac{1}{3}\right)\left(\frac{1}{3}\right) = \frac{1}{9}.$$

To calculate $\Pr\{X_n = j\}$ for larger n, it is useful to observe that

$$\Pr\{X_{n+1} = j\} = \sum_i p_{ij} \Pr\{X_n = i\}. \tag{3}$$

The only way to get to state j at time $n + 1$ is to be in some state i at time n, and then jump from i to j. Hence, we could have calculated

$$\Pr\{X_2 = 1\} = p_{11} \Pr\{X_1 = 1\} + p_{21} \Pr\{X_1 = 2\} + p_{31} \Pr\{X_1 = 3\},$$

and so forth. This is where the matrix notation comes in handy.

If we let

$$\pi_n(i) = \Pr\{X_n = i\},$$

then Eq. (3) can be written in the form

$$\pi_{n+1}(j) = \sum_i p_{ij}\, \pi_n(i). \tag{4}$$

If we let π_n denote the vector with entries $\pi_n(1)$, $\pi_n(2)$, ..., and let P denote the matrix in Eq. (2), then the set of equations relating π_{n+1} to π_n can be written most compactly in the form

$$\pi_{n+1} = \pi_n P. \tag{5}$$

For example, we have that $\pi_2 = \pi_1 P$, or

$$(0.67\bar{7},\ 0.21\bar{1},\ \tfrac{1}{9}) = \left(\tfrac{1}{3},\ \tfrac{1}{3},\ \tfrac{1}{3}\right) \begin{pmatrix} 1/3 & 1/3 & 1/3 \\ 0.7 & 0.3 & 0 \\ 1 & 0 & 0 \end{pmatrix}.$$

Now we can calculate $\pi_3 = \pi_2 P$ to get

$$\pi_3 = (0.485,\ 0.289,\ 0.226)$$

to three decimal places. Continuing on, we obtain

$$\pi_4 = (0.590,\ 0.248,\ 0.162)$$
$$\pi_5 = (0.532,\ 0.271,\ 0.197)$$
$$\pi_6 = (0.564,\ 0.259,\ 0.177)$$
$$\pi_7 = (0.546,\ 0.266,\ 0.188)$$
$$\pi_8 = (0.556,\ 0.262,\ 0.182)$$
$$\pi_9 = (0.551,\ 0.264,\ 0.185)$$
$$\pi_{10} = (0.553,\ 0.263,\ 0.184)$$
$$\pi_{11} = (0.553,\ 0.263,\ 0.184)$$
$$\pi_{12} = (0.553,\ 0.263,\ 0.184).$$

Notice that the probabilities $\pi_n(i) = \Pr\{X_n = i\}$ tend to a specific limiting value as n increases. We say that the stochastic process approaches steady state. This concept of steady state or equilibrium differs from that for a deterministic dynamic model. Because of random fluctuations, we cannot expect that the state variable will stay at one value when the system is in equilibrium. The best we can hope for is that the probability distribution of the state variable will tend to a limiting distribution. We call this the *steady-state distribution*. In Example 8.2 we have

$$\pi_n \rightarrow \pi,$$

where the steady-state probability vector

$$\pi = (0.553,\ 0.263,\ 0.184) \tag{6}$$

to three decimal places.

A faster way to calculate the steady-state vector π is as follows. Suppose that

$$\pi_n \rightarrow \pi.$$

Certainly

$$\pi_{n+1} \rightarrow \pi$$

too, so if we let $n \rightarrow \infty$ on both sides of Eq. (5), we obtain the equation

$$\pi = \pi P. \tag{7}$$

We can calculate π simply by solving this linear system of equations. For Example 8.2 we have

$$(\pi_1, \pi_2, \pi_3) = (\pi_1, \pi_2, \pi_3) \begin{pmatrix} \frac{1}{3} & \frac{1}{3} & \frac{1}{3} \\ 0.7 & 0.3 & 0 \\ 1 & 0 & 0 \end{pmatrix},$$

and it is not hard to calculate that Eq. (6) is the only solution to this system of equations for which

$$\sum \pi_i = 1.$$

Not every Markov chain tends to steady state. For example, consider the two-state Markov chain for which

$$\Pr\{X_{n+1} = 2 | X_n = 1\} = 1$$

and

$$\Pr\{X_{n+1} = 1 | X_n = 2\} = 1.$$

The state variable alternates between states 1 and 2. Certainly π_n does not tend to a single limiting vector. We say that this Markov chain is periodic with period two. Generally we say X_n is periodic with period δ if, starting at $X_n = i$, the chain can return to state i only at times $n + k\delta$. If $\{X_n\}$ is aperiodic ($\delta = 1$), and if for each i and j it is possible to transition from i to j in a finite number of steps, we say that X_n is *ergodic*. There is a theorem that guarantees that an ergodic Markov chain tends to steady state. Furthermore, the distribution of X_n tends to the same steady-state distribution regardless of the initial state of the system (see, for example, Çinlar, E., (1975), p. 152). Thus, in Example 8.2, if we had started with $X_0 = 2$ or $X_0 = 3$, we would still see π_n converge to the same steady-state distribution π given by Eq. (6). The problem of calculating the steady-state probability vector π is mathematically equivalent to the problem of locating the equilibrium of a discrete-time dynamical system with state space $\pi \in \mathbb{R}^m$, $0 \leq \pi_j \leq 1$,

$$\sum \pi_i = 1$$

and iteration function

$$\pi_{n+1} = \pi_n P.$$

The previously-cited theorem states that there is a unique asymptotically stable equilibrium π for this system whenever P represents an ergodic Markov chain.

We return now to the inventory problem of Example 8.1. We will model this problem using a Markov chain. Step 3 is to formulate the model. We begin with a consideration of the state space. The concept of state here is much the same as for deterministic dynamical systems. The state contains all of the information necessary in order to predict the (probabilistic) future of the process. We will take

$X_n = S_n$, the number of 20-gallon aquariums in stock at the beginning of our sales week, as the state variable. The demand D_n relates to the dynamics of the model and will be used to construct the state transition matrix P. The state space is

$$X_n \in \{1, 2, 3\}.$$

We do not know the initial state, but it seems reasonable to assume that $X_0 = 3$. In order to determine P we will begin by drawing the state transition diagram. See Figure 8.3. The distribution of the demand D_n yields

$$\Pr\{D_n = 0\} = 0.368$$
$$\Pr\{D_n = 1\} = 0.368$$
$$\Pr\{D_n = 2\} = 0.184 \tag{8}$$
$$\Pr\{D_n = 3\} = 0.061$$
$$\Pr\{D_n > 3\} = 0.019,$$

so that if $X_n = 3$, then

$$\Pr\{X_{n+1} = 1\} = \Pr\{D_{n+1} = 2\} = 0.184$$
$$\Pr\{X_{n+1} = 2\} = \Pr\{D_{n+1} = 1\} = 0.368$$
$$\Pr\{X_{n+1} = 3\} = 1 - (0.184 + 0.368) = 0.448.$$

The remaining state transition probabilities are computed similarly. The state

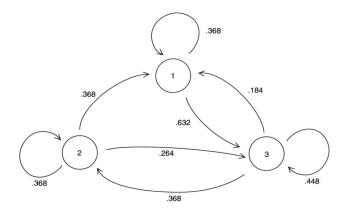

Figure 8.3 State transition diagram for the inventory problem.

transition matrix is

$$P = \begin{pmatrix} .368 & 0 & .632 \\ .368 & .368 & .264 \\ .184 & .368 & .448 \end{pmatrix}. \tag{9}$$

Now for step 4. The analysis objective was to calculate the probability

$$\Pr\{D_n > S_n\}$$

that demand exceeds supply. In general this probability depends on n. More specifically, it depends on X_n. If $X_n = 3$, then

$$\Pr\{D_n > S_n\} = \Pr\{D_n > 3\} = 0.019,$$

and so forth. To get a better idea of how often demand will exceed supply, we need to know more about X_n.

Since $\{X_n\}$ is an ergodic Markov chain, we know that there is a unique steady-state probability vector π that can be computed by solving the steady-state equations. Substituting Eq. (9) back into Eq. (7), we obtain

$$\begin{aligned} \pi_1 &= .368\pi_1 + .368\pi_2 + .184\pi_3 \\ \pi_2 &= .368\pi_2 + .368\pi_3 \\ \pi_3 &= .632\pi_1 + .264\pi_2 + .448\pi_3, \end{aligned} \tag{10}$$

which we need to solve along with the condition

$$\pi_1 + \pi_2 + \pi_3 = 1,$$

to obtain the steady-state distribution of X_n. Since we now have four equations in three variables, we can delete one of the equations in Eq. (10) and then solve, which yields

$$\pi = (\pi_1, \pi_2, \pi_3) = (.285, .263, .452).$$

For all large n it is approximately true that

$$\begin{aligned} \Pr\{X_n = 1\} &= .285 \\ \Pr\{X_n = 2\} &= .263 \\ \Pr\{X_n = 3\} &= .452. \end{aligned}$$

Putting this together with our information about D_n, we obtain

$$\Pr\{D_n > S_n\} = \sum_{i=1}^{3} \Pr\{D_n > S_n | X_n = i\} \Pr\{X_n = i\}$$

$$= (.264)(.285) + (.080)(.263) + (.019)(.452) = .105$$

```
> s:={pi1=.368*pi1+.368*pi2+.184*pi3,
>       pi2=.368*pi2+.368*pi3,
>       pi1+pi2+pi3=1};
   s:={pi1 = .368 pi1 + .368 pi2 + .184 pi3, pi2 = .368 pi2 + .368 pi3,

         pi1 + pi2 + pi3 = 1}

> solve(s,{pi1,pi2,pi3});
         {pi3 = .4519843569, pi1 = .2848348783, pi2 = .2631807648}

> stop;
```

Figure 8.4 Calculation of the steady-state distribution of the number of 20-gallon aquariums in stock at the beginning of the week for the inventory problem, using the computer algebra system MAPLE.

for large n. In the long run, demand will exceed supply about 10% of the time.

It is easy to compute the steady-state probabilities using a computer algebra system. Figure 8.4 illustrates the use of the computer algebra system MAPLE to solve the system of equations in Eq. (10) to find the steady-state probabilities. Computer algebra systems are quite useful in such problems, especially when performing sensitivity analysis. If you have access to a computer algebra system, it will be useful in the exercises at the end of this chapter. Even if you prefer to solve systems of equations by hand, you will have the ability to verify your results.

Finally, step 5. The current inventory policy results in lost sales about 10% of the time, or at least five lost sales per year. Most of this is due to the fact that we do not order more aquariums when only one is left. Although we only sell an average of one unit per week, the actual number of potential sales per week (demand) fluctuates from one week to the next. Hence when we start the week with only one unit in stock, we run a significant risk (about a 1 in 4 chance) of losing potential sales due to insufficient inventory. In the absence of other factors, such as a discount for orders of three or more, it seems reasonable to try out a new inventory policy in which we never start out a week with only one aquarium.

We now come to the subject of sensitivity analysis and robustness. The main sensitivity issue is the effect of the arrival rate λ of potential buyers on the probability that demand exceeds supply. Currently $\lambda = 1$ customer per week. For arbitrary λ, the state transition matrix for X_n is given by

$$P = \begin{pmatrix} e^{-\lambda} & 0 & 1 - e^{-\lambda} \\ \lambda e^{-\lambda} & e^{-\lambda} & 1 - (1 + \lambda)e^{-\lambda} \\ \lambda^2 e^{-\lambda}/2 & \lambda e^{-\lambda} & 1 - (\lambda + \lambda^2/2)e^{-\lambda} \end{pmatrix}, \tag{11}$$

using the fact that D_n has a Poisson distribution. While it would be possible to carry through the calculation of $p = \Pr\{D_n > S_n\}$ from this point, it would be very messy. It makes more sense simply to repeat the calculations of step 4 for a few selected values of λ near 1. The results of this exercise are shown in Figure 8.5. They confirm that our basic conclusions are not particularly sensitive to the exact value of λ. The sensitivity $S(p, \lambda)$ is around 1.5. (Another sensible option for sensitivity analysis is to use a computer algebra system to perform the messy calculations. See Exercise 2 at the end of this chapter.)

Finally, we should consider the robustness of our model. We have assumed a Markov chain model based on a Poisson process model of the arrival process. The robustness of the Poisson process model as a representative of a more general arrival process was discussed briefly at the end of Section 7.2. It is reasonable to conclude that our results would not be altered significantly if the arrival process were not exactly Poisson. The basic assumption here is that the arrival process represents the merging of a large number of independent arrival processes. Many kinds of customers arrive at the shop from time to time to buy a 20-gallon aquarium, and it is reasonable to assume that they do not coordinate this activity with each other. Of course, certain store activities such as an advertised sale price on 20-gallon aquariums, would invalidate this assumption, resulting in the need to reexamine the conclusions of our modeling exercise. It may also be that there are significant seasonal variations in the demand for this item.

The other basic modeling assumption is that the inventory level S_n represents

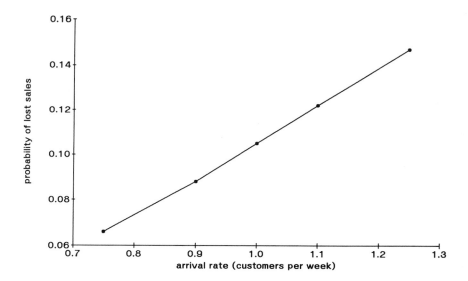

Figure 8.5 Graph showing the sensitivity of the probability of lost sales to arrival rate in the inventory problem.

the state of the system. A more sophisticated model might take into account the store manager's response to long-term fluctuations in sales, such as seasonal variations. The mathematical analysis of such a model is more complex, but not essentially different from what we did here. We just expand the state space to include information on past sales, say, S_n, S_{n-1}, S_{n-2}, S_{n-3}. Of course, our transition matrix P is now 81×81 instead of 3×3.

Many different inventory policies are possible. Several of these are explored in the exercises at the end of the chapter. Which inventory policy is the best? One way to approach this question is to formulate an optimization model based on a generalized version of our Markov chain model. A range of inventory policies is described in terms of one or more decision variables, and the objective is defined in terms of the resulting steady-state probabilities. The study of such models is called *Markov decision theory*. Details can be found in any introductory text on operations research (e.g.. Hillier, F. et al., (1990)).

8.2   Markov Processes

A Markov process model is the continuous-time analogue of the Markov chain model introduced in the preceding section. It may also be considered as a stochastic analogue to a continuous-time dynamical system model.

Example 8.3   A mechanic working for a warehouse distribution center is responsible for the repair and maintenance of the trucks used there. When forklifts break down they are taken to a repair facility and serviced in the order of their arrival. There are 27 forklift trucks in use at the warehouse, and an average vehicle needs repair about once every six months. Average repair time for a single vehicle is about three days. In the past few months certain questions have been raised about the effectiveness and the efficiency of this operation. The two central issues are the time it takes to have a machine repaired, and the percentage of time the mechanic devotes to this part of his duties.

We will analyze the situation using a mathematical model of the repair facility. Forklift trucks arrive at the service facility for repair at a rate of $27/6 = 4.5$ per month. The maximum rate at which they can be repaired is $22/3 \approx 7.3$ vehicles per month, based on an average of 22 working days per month. Let X_t denote the number of vehicles in the repair shop at time t. We are interested in the average number in service, EX_t, and the proportion of time that the mechanic is busy repairing machines, represented by $\Pr\{X_t > 0\}$. Figure 8.6 summarizes the results of step 1.

We will model the repair facility using a Markov process.

Variables:	X_t = number of forklifts in repair at time t months.
Assumptions:	Vehicles arrive for repair at the rate of 4.5 per month. Maximum repair rate is 7.3 vehicles per month.
Objective:	Calculate EX_t, $\Pr\{X_t > 0\}$

Figure 8.6 Results of step 1 of the forklift problem.

A *Markov process* is the continuous-time analogue of the Markov chain introduced in the previous section. As before we will assume that the state space is finite, i.e., we will suppose

$$X_t \in \{1, 2, 3, \ldots, m\}.$$

The stochastic process $\{X_t\}$ is a Markov process if the current state X_t really represents the state of the system, i.e., it totally determines the probabilistic future of the process. This condition is formally written as

$$\Pr\{X_{t+s} = j \,|\, X_u : u \leq t\} = \Pr\{X_{t+s} = j \,|\, X_t\}. \tag{12}$$

The Markov property, Eq. (12), has two important implications. First of all, the time until the next transition does not depend on how long the process has been in the current state. In other words, the distribution of time spent in a particular state has the memoryless property. Let T_i denote the time spent in state i. Then the Markov property says that

$$\Pr\{T_i > t + s \,|\, T_i > s\} = \Pr\{T_i > t\}. \tag{13}$$

In Section 7.2 we showed that the negative exponential distribution has this property, so T_i could have density function

$$F_i(t) = \lambda_i e^{-\lambda_i t}. \tag{14}$$

In fact, the negative exponential distribution is the only probability distribution having the memoryless property. (This is a deep theorem in real analysis. See Billingsley, P., (1979), p. 160.) Hence, for a Markov process the distribution of time in a particular state is negative exponential with parameter λ_i, which depends in general on the state i.

The second important implication of the Markov property has to do with state transition. The probability distribution that describes the identity of the next state can depend only on the current state. Thus, the sequence of states visited by the process forms a Markov chain. If we let p_{ij} denote the probability that the process jumps from state i to state j, then the embedded Markov chain has state transition probability matrix $P = (p_{ij})$.

Example 8.4 Consider a Markov chain with state transition probability

$$P = \begin{pmatrix} 0 & 1/3 & 2/3 \\ 1/2 & 0 & 1/2 \\ 3/4 & 1/4 & 0 \end{pmatrix}, \tag{15}$$

and form a Markov process by assuming that the jumps of $\{X_t\}$ follow this Markov chain with the average time in state 1, 2, 3 equal to 1, 2, and 3, respectively.

Solving the steady-state equation $\pi = \pi P$ shows that the proportion of jumps that land in states 1, 2, and 3 are 0.396, 0.227, and 0.377, respectively. However, the proportion of time spent in each state also depends on how long we wait in one state before the next jump. Correcting for this produces the relative proportions 1(.396), 2(.227), and 3(.377). If we normalize to 1 (divide each term by the sum), we get .200, .229, and .571. Hence the Markov process spends about 57.1% of the time in state 3, and so on. We call this the steady-state distribution for the Markov process. Generally, if $\pi = (\pi_1, \ldots, \pi_m)$ is the steady-state distribution for the embedded Markov chain and $\lambda = (\lambda_1, \ldots, \lambda_m)$ is the vector of rates, then the proportion of time spent in state i is given by

$$P_i = \frac{(\pi_i/\lambda_i)}{(\pi_1/\lambda_1) + \cdots + (\pi_m/\lambda_m)}. \tag{16}$$

The reciprocal of the rate λ_i represents mean time in state i. In summary, a Markov process can be thought of as a Markov chain where the time between jumps has a negative exponential distribution, depending on the current state. An equivalent model can be formulated as follows. Given that $X_t = i$, let T_{ij} be negative exponential with parameter $a_{ij} = \lambda_i p_{ij}$. Furthermore, suppose that $T_{i1}, \ldots, T_{im}$ are independent. Then the time T_i until the next jump is the minimum of $T_{i1}, \ldots, T_{im}$, and the next state is the state j such that T_{ij} is the minimum of $T_{i1}, \ldots, T_{im}$. The mathematical equivalence between the two forms of the Markov process model follows from the fact that

$$T_i = \min(T_{i1}, \ldots, T_{im})$$

is negative exponential with parameter

$$\lambda_i = \sum_j a_{ij}$$

and that

$$\Pr\{T_i = T_{ij}\} = p_{ij}.$$

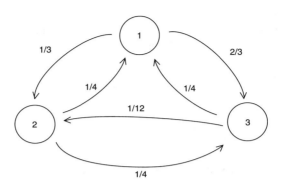

Figure 8.7 Rate diagram for Example 8.4, illustrating the rate at which the process tends to jump from one state to another.

(The proof is left to the reader. See Exercise 6 at the end of the chapter.) The parameter

$$a_{ij} = \lambda_i p_{ij}$$

denotes the rate at which the process tends to go from state i to state j. It is customary to depict the rates a_{ij} in a *rate diagram*. The rate diagram for Example 8.4 is given in Figure 8.7. Often the structure of a Markov process is originally specified by such a diagram.

This alternative formulation for the Markov process leads to a convenient method of computing the steady-state distribution, based on the rate diagram. As before we define

$$a_{ij} = \lambda_i p_{ij}$$

to be the rate at which the process tends to jump from state i to state j. Also let

$$a_{ii} = -\lambda_i$$

denote the rate at which the process tends to leave state i. It can be shown that the probability functions

$$P_i(t) = \Pr\{X_t = i\} \tag{17}$$

must satisfy the differential equations

$$P_1'(t) = a_{11} P_1(t) + \cdots + a_{1m} P_m(t)$$

$$\vdots \tag{18}$$

$$P_m'(t) = a_{m1} P_1(t) + \cdots + a_{mm} P_m(t)$$

(see Çinlar, E., (1975), p. 255). This basic condition can be most easily understood by using the *fluid-flow analogy*. Visualize the probabilities $P_i(t)$

as the amount of fluid (probability mass) at each state i. The rates a_{ij} represent the rate of fluid flow, and the fact that

$$P_1(t) + \cdots + P_m(t) = 1$$

means that the total amount of fluid stays equal to 1. In the case of Example 8.4 we have

$$
\begin{aligned}
P_1'(t) &= -P_1(t) + 1/4 \ P_2(t) + 1/4 \ P_3(t) \\
P_2'(t) &= 1/3 \ P_1(t) - 1/2 \ P_2(t) + 1/12 \ P_3(t) \\
P_3'(t) &= 2/3 \ P_1(t) + 1/4 \ P_2(t) - 1/3 \ P_3(t).
\end{aligned}
\tag{19}
$$

The steady-state distribution for the Markov process corresponds to the steady-state solution to this system of differential equations. Setting $P_i' = 0$ for all i, we obtain

$$
\begin{aligned}
0 &= -P_1 + 1/4 \ P_2 + 1/4 \ P_3 \\
0 &= 1/3 \ P_1 - 1/2 \ P_2 + 1/12 \ P_3 \\
0 &= 2/3 \ P_1 + 1/4 \ P_2 - 1/3 \ P_3.
\end{aligned}
\tag{20}
$$

Solving the system of linear equations in Eq. (20) together with the condition

$$P_1 + P_2 + P_3 = 1$$

yields

$$P = (7/35, \ 8/35, \ 20/35),$$

which is the same as before.

One simple way to determine the system of equations which must be solved to get the steady-state distribution for a Markov process is to use the fluid-flow analogy. Fluid flows into and out of each state. In order that the system remain in equilibrium, the rate at which fluid flows into each state must equal the rate at which it flows back out. For example, in Fig. 8.7 we have that fluid is flowing out of state 1 at the rate of $1/3 + 2/3 = 1 \times P_1$. Fluid flows from state 2 back into state 1 at the rate of $1/4 \times P_2$ and from state 3 to state 1 at the rate of $1/4 \times P_3$, so we have the condition

$$P_1 = 1/4 \ P_2 + 1/4 \ P_3.$$

By applying this principle of

$$[\text{Rate out}] \ = [\text{Rate in}]$$

to the other two states as well, we obtain the system of equations

$$
\begin{aligned}
P_1 &= 1/4 \ P_2 + 1/4 \ P_3 \\
1/2 \ P_2 &= 1/3 \ P_1 + 1/12 \ P_3 \\
1/3 \ P_3 &= 2/3 \ P_1 + 1/4 \ P_2.
\end{aligned}
\tag{21}
$$

which is equivalent to the system of equations in Eq. (20). We call the system of equations in Eq. (21) the *balance equations* for our Markov process model. They express the condition that the rates into and out of each state are in balance.

In Section 8.1 we remarked that an ergodic Markov chain will always tend to steady state. Now we will state the corresponding result for Markov processes. A Markov process is called *ergodic* if for every pair of states i and j it is possible to jump from i to j in a finite number of transitions. There is a theorem that guarantees that an ergodic Markov process always tends to steady state. Furthermore, the distribution of X_t tends to the same steady-state distribution regardless of the initial state of the system (see, for example, Çinlar, E., (1975), p. 264). Let

$$P(t) = (P_1(t), \ldots, P_m(t))$$

denote the current probability distribution of our Markov process. Then this theorem says that for any initial probability distribution $P(0)$ on the state space, we will always see the probability distribution $P(t)$ of the Markov process state vector X_t converge to the same steady-state distribution

$$P = (P_1, \ldots, P_m)$$

as $t \to \infty$. The system of differential equations in Eq. (18) which describes the dynamics of the probability distribution $P(t)$ can be written in matrix form as

$$P(t)' = P(t)A, \tag{22}$$

where $A = (a_{ij})$ is the matrix of rates. This is a linear system of differential equations on the space

$$S = \{x \in \mathbb{R}^m : 0 \le x_i \le 1; \sum x_i = 1\}. \tag{23}$$

Our theorem says that if this dynamical system in Eq. (22) represents an ergodic Markov process, then there exists a unique stable equilibrium solution P. Furthermore, for any initial condition $P(0)$ we will have $P(t) \to P$ as $t \to \infty$. More detailed information about the transient (time-dependent) behavior of the Markov process can be obtained by explicitly solving the linear system in Eq. (22) by the usual methods.

Now we return to the forklift problem of Example 8.3. We want to formulate a Markov process model for X_t = the number of forklifts in repair at time t months. Since there are only 27 forklifts we have

$$X_t \in \{0, 1, 2, \ldots, 27\}.$$

The only allowed transitions are from $X_t = i$ to $X_t = i + 1$ or $i - 1$. The rates up and down are $\lambda = 4.5$ and $\mu = 7.3$, respectively, except that we cannot transition up from state 27 or down from state 0. The rate diagram for this problem is shown in Figure 8.8.

The steady-state equations $PA = 0$ can be obtained from the rate diagram by using the

$$[\text{Rate out}] = [\text{Rate in}]$$

principle. From Fig. 8.8 we obtain

$$\lambda P_0 = \mu P_1$$
$$(\mu + \lambda) P_1 = \lambda P_0 + \mu P_2$$
$$(\mu + \lambda) P_2 = \lambda P_1 + \mu P_3$$
$$\vdots$$
$$(\mu + \lambda) P_{26} = \lambda P_{25} + \mu P_{27}$$
$$\mu P_{27} = \lambda P_{26}.$$

(24)

Solving along with

$$\sum P_i = 1$$

will yield the steady-state $\Pr\{X_t = i\}$. We are interested in

$$\Pr\{X_t > 0\} = 1 - \Pr\{X_t = 0\} = 1 - P_0$$

and in

$$EX_t = \sum i P_i.$$

Moving on to step 4, we first solve for P_1 in terms of P_0, then for P_2 in terms of P_1, and so forth, to obtain

$$P_n = (\lambda/u) P_{n-1}$$

for all $n = 1, 2, 3, \ldots, 27$. Then

$$P_n = (\lambda/u)^n P_0$$

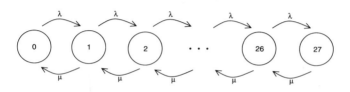

Figure 8.8 Rate diagram for the forklift problem, illustrating the rate at which the number of forklifts in service tends to increase or decrease.

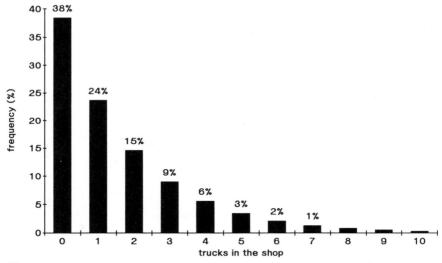

Figure 8.9 Histogram showing the distribution of the number of forklifts in service at any given time.

for all such n. Since

$$\sum_{n=0}^{27} P_n = P_0 \sum_{n=0}^{27} \left(\frac{\lambda}{u}\right)^n = 1,$$

we must have

$$P_0 = (1 - \rho)/(1 - \rho^{28}),$$

where $\rho = \lambda/u$. Here we have used the standard formula for the sum of a finite geometric series. For $n = 1, 2, 3, \ldots, 27$ we have

$$P_n = \rho^n P_0 = \frac{\rho^n (1 - \rho)}{1 - \rho^{28}}. \tag{25}$$

Now $\rho = \lambda/u = 4.5/7.3 \approx 0.616$, so $1 - \rho^{28} \approx 0.9999987$. Hence, we may assume for all practical purposes that $P_0 = 1 - \rho$ and $P_n = \rho^n(1 - \rho)$ for $n \geq 1$.

Now we calculate our two measures of performance. First we have

$$\Pr\{X_t > 0\} = 1 - P_0 = \rho \approx 0.616.$$

Next we have

$$EX_t = \sum_{n=0}^{27} n P_n$$

$$= \sum_{n=0}^{27} n\rho^n (1 - \rho), \tag{26}$$

which yields $EX_t = 1.607$.

To summarize (step 5) we consider a system where forklift trucks break down at a rate of 4.5 per month and are taken to a repair facility with the capacity to service up to 7.3 per month. Since the rate at which vehicles arrive for repair is only about 60% of the potential service rate, the mechanic is busy with this activity only about 60% of the time. However, since breakdowns occur essentially at random, there will be times when there is more than one vehicle in the shop at one time, through no fault of the mechanic. In fact, on an average day we would expect to see 1.6 vehicles in the repair facility. By this we mean that if we kept track on a daily basis of the number of vehicles in the repair facility, then at the end of the year this number would average about 1.6. More specifically, we would expect the distribution indicated by Fig. 8.9.

Murphy's law is certainly in evidence here. Even if the mechanic does his job flawlessly, on 8% of the working days in a year there will be 5 or more forklift trucks in the shop. Since it takes three days to fix one truck, this represents a backlog of about three weeks. Assuming 250 working days a year, this unfortunate situation will occur about 20 days out of the year. Meanwhile, a time study will show the mechanic busy on this part of his job only about 60% of the time. This seeming discrepancy can be explained by the simple fact that breakdowns are not always nicely spaced out. Sometimes, due to sheer bad luck, several machines will break down in rapid succession, and the mechanic will be swamped. At other times there will be long periods between breakdowns, and the mechanic will have no repairs to perform.

There are really two management problems here. One is the problem of idle time. There is no way to avoid the sporadic nature of this job activity, so idle time will continue to be a significant concern. Management might consider the extent to which the mechanic can perform other duties during this "idle time." Of course, these duties should be easily interruptible in case there is a breakdown.

The second problem is backlogs. Sometimes a number of forklift trucks will be out of service. Additional study may be required to determine whether this problem should be addressed by purchasing an additional reserve of forklift trucks or whether additional manpower should be devoted to the repair activity during particularly busy periods.

Finally we come to the very important questions of sensitivity and robustness. Both measures of performance depend on the ratio

$$\rho = \frac{\lambda}{\mu},$$

currently equal to 0.616. The relation

$$\Pr\{X_t > 0\} \approx \rho \qquad (27)$$

requires no further comment. If we let $A = EX_t$ denote the average number in the system, then (using $1 - \rho^{28} \approx 1$) we have

$$A = \sum_{n=0}^{27} n\rho^n(1 - \rho), \qquad (28)$$

which reduces to

$$A = \rho + \rho^2 + \rho^3 + \cdots + \rho^{27} - 27\rho^{28}.$$

This is approximately

$$\rho(1 + \rho + \rho^2 + \rho^3 + \cdots) = \rho/(1 - \rho),$$

and so

$$\frac{dA}{d\rho} \approx \frac{1}{(1 - \rho)^2},$$

so that

$$S(A, \rho) \approx 2.6.$$

A small error in ρ would not significantly alter our conclusions based on $A = EX_t$.

We should also consider the size of the operational fleet, currently at $K = 27$ forklift trucks. We have seen that for moderate ρ (not too close to 1) this parameter makes little difference, in the sense that we have continually used the approximation

$$1 - \rho^{K+1} \approx 1.$$

Indeed this approximation is equivalent to the assumption that, from the mechanic's point of view, there is an unlimited supply of forklift trucks to break down. Essentially we are assuming that $K = \infty$. If more vehicles are purchased, this will not change the rate λ at which vehicles break down (we are only buying more, not using more). Thus $\rho = \lambda/\mu$ remains the same, so both our measures of performance are insensitive to K. The only effect of increasing K (i.e., buying additional trucks) will be to increase the number of trucks that remain operational.

The model we have used to represent the repair facility is a special case of a *queuing model*. A queuing model represents a system consists of one or more service facilities at which arrivals are processed, and those arrivals that cannot be processed have to wait for service in a queue. There is a large body of literature on queuing models, including much current research. A textbook on operations research (e.g., Hillier, F. et al., (1990)) is a good place to begin. The most important assumption we might try to relax is that service times have a negative exponential distribution. There is ample reason to suspect that the arrivals are more or less random. There is a result (based on the simplifying assumption $K = \infty$, which we have made before) for a general service time distribution with variance σ^2 that states that $\rho = \lambda/\mu$ is the probability that the server is busy and the steady-state

$$EX_t = \frac{\rho + \lambda^2 \sigma^2 + \rho^2}{2(1 - \rho)}. \tag{29}$$

Of course this reduces to $\rho/(1 - \rho)$ in the case of a negative exponential service time. In that case, $\sigma = 1/\mu$. The general conclusion to be drawn from this formula is that the average number of vehicles in repair grows with the variance of service time. Thus, more uncertainty about the length of a repair will result in longer waiting times.

8.3   Linear Regression

The single most commonly used stochastic model assumes that the expected value of the state variable is a linear function of time. The model is attractive not only because of its wide range of applications, but also because of the availability of good software implementations.

Example 8.5   Adjustable-rate mortgages on private homes are commonly based on one of several market indices tabulated by the federal home loan bank. The author's mortgage is adjusted yearly on the basis of the one-year U.S. Treasury Constant Maturity index (CM1) for May of each year. Historical data for the three-year period beginning June 1986 are shown in Table I (source: Board of Governors of the Federal Reserve). Use this information to project the estimated value of this index in May 1990, the date of the next adjustment.

We will use the five-step method. Step 1 is to ask a question. We are attempting to estimate the future trend in a variable that exhibits a tendency to grow with time, along with some random fluctuation. Let X_t denote the one-year U.S. Treasury Constant Maturity index (CM1) at time t months after May 1986. A graph of X_t for $t = 1, \ldots, 37$ is shown in Figure 8.10. We want to estimate X_{48}. If we assume that X_t depends in part on a random element, then we cannot expect to predict X_{48} exactly. The best we can hope for is an average value EX_{48}, along with some kind of measure of the magnitude of uncertainty. For the moment we will concentrate on obtaining an estimate of EX_{48}. We will leave the other matter for the section on sensitivity analysis.

Step 2 is to select the modeling approach. We will model this problem using linear regression.

Table I.   Possible adjustable-rate mortgage loan indices.

	TB3	TB6	CM1	CM2	CM3	CM5
6/86	6.21	6.28	6.73	7.18	7.41	7.64
7/86	5.84	5.85	6.27	6.67	6.86	7.06
8/86	5.57	5.58	5.93	6.33	6.49	6.80
9/86	5.19	5.31	5.77	6.35	6.62	6.92
10/86	5.18	5.26	5.72	6.28	6.56	6.83
11/86	5.35	5.42	5.80	6.28	6.46	6.76
12/86	5.49	5.53	5.87	6.27	6.43	6.67
1/87	5.45	5.47	5.78	6.23	6.41	6.64
2/87	5.59	5.60	5.96	6.40	6.56	6.79
3/87	5.56	5.56	6.03	6.42	6.58	6.79
4/87	5.76	5.93	6.50	7.02	7.32	7.57
5/87	5.75	6.11	7.00	7.76	8.02	8.26
6/87	5.69	5.99	6.80	7.57	7.82	8.02
7/87	5.78	5.86	6.68	7.44	7.74	8.01
8/87	6.00	6.14	7.03	7.75	8.03	8.32
9/87	6.32	6.57	7.67	8.34	8.67	8.94
10/87	6.40	6.86	7.59	8.40	8.75	9.08
11/87	5.81	6.23	6.96	7.69	7.99	8.35
12/87	5.80	6.36	7.17	7.86	8.13	8.45
1/88	5.90	6.31	6.99	7.63	7.87	8.18
2/88	5.69	5.96	6.64	7.18	7.38	7.71
3/88	5.69	5.91	6.71	7.27	7.50	7.83
4/88	5.92	6.21	7.01	7.59	7.83	8.19
5/88	6.27	6.53	7.40	8.00	8.24	8.58
6/88	6.50	6.76	7.49	8.03	8.22	8.49
7/88	6.73	6.97	7.75	8.28	8.44	8.66
8/88	7.02	7.36	8.17	8.63	8.77	8.94
9/88	7.23	7.43	8.09	8.46	8.57	8.69
10/88	7.34	7.50	8.11	8.35	8.43	8.51
11/88	7.68	7.76	8.48	8.67	8.72	8.79
12/88	8.09	8.24	8.99	9.09	9.11	9.09
1/89	8.29	8.38	9.05	9.18	9.20	9.15
2/89	8.48	8.49	9.25	9.37	9.32	9.27
3/89	8.83	8.87	9.57	9.68	9.61	9.51
4/89	8.70	8.73	9.36	9.45	9.40	9.30
5/89	8.40	8.39	8.98	9.02	8.98	8.91
6/89	8.22	8.00	8.44	8.41	8.37	8.29

The *linear regression* model assumes that

$$X_t = a + bt + \varepsilon_t, \tag{30}$$

where a and b are real constants and ε_t is a random variable that represents the effect of random fluctuations. It is assumed that

$$\varepsilon_1, \ \varepsilon_2, \ \varepsilon_3, \ \ldots$$

are independent and identically distributed with mean zero. It is also common to assume that ε_t is normal, i.e., that for some $\sigma > 0$ the random variable

$$\varepsilon_t / \sigma$$

has the standard normal density. In the case where the random fluctuations represented by ε_t involve the additive effects of a fairly large number of independent random factors, this normal assumption is justified by the central limit theorem. (The normal density and the central limit theorem were introduced in Section 7.3.)

Since the error term ε_t has mean zero,

$$E X_t = a + bt, \tag{31}$$

so the problem of estimating $E X_t$ reduces to the estimation of the parameters a and b. If we were to graph the line

$$y = a + bt \tag{32}$$

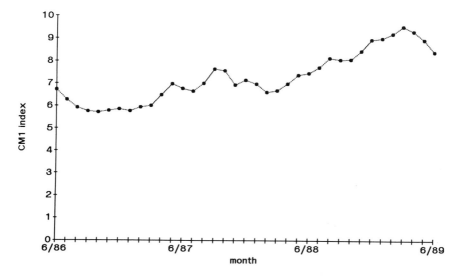

Figure 8.10 Graph of CM1 index versus time for the ARM problem.

on the graph of Fig. 8.10, we would expect the data points (the corners on the CM1 graph) to lie near this line, with some above and some below. The best-fitting line, representing our best estimate of the parameters a and b, should minimize the extent to which the data points deviate from the line.

Given a set of data points

$$(t_1, y_1), \ldots, (t_n, y_n),$$

we measure the goodness of fit of the regression line in terms of the vertical distance

$$|y_i - (a + bt_i)|$$

between the data point (t_i, y_i) and the point on the regression line, Eq. (32), at $t = t_i$. To avoid absolute value signs, which are troublesome in an optimization problem, we measure the overall goodness of fit by

$$F(a, b) = \sum_{i=1}^{n} (y_i - (a + bt_i))^2. \tag{33}$$

The best fitting line is characterized by a global minimum of the objective function in Eq. (33). Setting the partial derivatives $\partial F / \partial a$ and $\partial F / \partial b$ equal to zero yields

$$\sum_{i=1}^{n} y_i = na + b \sum_{i=1}^{n} t_i$$

$$\sum_{i=1}^{n} t_i y_i = a \sum_{i=1}^{n} t_i + b \sum_{i=1}^{n} t_i^2. \tag{34}$$

Solving these two linear equations in two unknowns determines a and b.

An estimate of the predictive power of the regression equation (32) can be obtained as follows. Let

$$\bar{y} = \frac{1}{n} \sum_{i=1}^{n} y_i \tag{35}$$

denote the average or mean value of the y data points, and for each i let

$$\hat{y}_i = a + bt_i. \tag{36}$$

The total variation $y_i - \bar{y}$ between one data value and the mean value can be expressed as the sum

$$(y_i - \bar{y}) = (y_i - \hat{y}_i) + (\hat{y}_i - \bar{y}). \tag{37}$$

The first term on the right-hand side of Eq. (37) represents the error (vertical distance of the data point from the regression line), and the second term represents the amount of deviation in y accounted for by the regression line. A little algebra shows that

$$\sum_{i=1}^{n} (y_i - \bar{y})^2 = \sum_{i=1}^{n} (y_i - \hat{y}_i)^2 + \sum_{i=1}^{n} (\hat{y}_i - \bar{y})^2. \tag{38}$$

The statistic

$$R^2 = \frac{\sum_{i=1}^{n} (\hat{y}_i - \bar{y})^2}{\sum_{i=1}^{n} (y_i - \bar{y})^2} \tag{39}$$

measures the portion of the total variation in the data accounted for by the regression line. The remaining portion of the total variation is attributed to random errors, i.e., the effect of ε_t. If R^2 is close to 1, then the data are very nearly linear. If R^2 is close to 0, the data are very nearly random.

Most mainframe computer installations have statistical packages that will automatically compute a, b, and R^2 from a data set. Inexpensive software of the same kind is available for most personal computers, and some hand-held calculators have built-in linear regression functions. For the linear regression problems in this book, any of these methods will suffice. It is not recommended that these problems be solved by hand.

Step 3 is to formulate the model. We will let X_t represent the value of the CM1 index t months after May 1986, and we will assume the linear regression model in Eq. (30). The data are

$$(t_1, y_1) = (1, 6.73)$$
$$(t_2, y_2) = (2, 6.27)$$
$$\vdots \tag{40}$$
$$(t_{37}, y_{37}) = (37, 8.44).$$

The best-fitting regression line can be obtained by solving the linear system of equations in Eq. (34) to obtain a and b. Then the goodness-of-fit statistic R^2 can be obtained from Eq. (39). Using a computer implementation of this linear regression model will allow us to avoid a lot of tedious calculation.

Step 4 is to solve the model. We used the MINITAB statistical package to obtain the regression line

$$y = 5.45 + 0.0970t \tag{41}$$

and

$$R^2 = 83.0\%$$

(see Figure 8.11). Equation (41) represents the best-fitting straight line through the data points in Eq. (40). See Figure 8.12 for a graphical illustration. The average trend in the CM1 index over the period June 1986 to June 1989 has been

```
set c1
1:37
end
set c2
6.73 6.27 5.93 5.77 5.72 5.80 5.87 5.78 5.96 6.03 6.50 7.00
6.80 6.68 7.03 7.67 7.59 6.96 7.17 6.99 6.64 6.71 7.01 7.40
7.49 7.75 8.17 8.09 8.11 8.48 8.99 9.05 9.25 9.57 9.36 8.98 8.44
end
regress c2 on 1 predictor c1;
predict 48.

The regression equation is
C2 = 5.45 + 0.0970 C1

Predictor          Coef        Stdev      t-ratio           p
Constant         5.4475       0.1615        33.73       0.000
C1              0.096989     0.007409        13.09       0.000

s = 0.4812        R-sq = 83.0%      R-sq(adj) = 82.6%

Analysis of Variance

SOURCE           DF            SS           MS          F          p
Regression        1        39.678       39.678     171.35      0.000
Error            35         8.104        0.232
Total            36        47.783

Unusual Observations
Obs.       C1           C2        Fit Stdev.Fit   Residual    St.Resid
   1      1.0       6.7300     5.5445    0.1551     1.1855       2.60R

R denotes an obs. with a large st. resid.


     Fit   Stdev.Fit          95% C.I.             95% P.I.
 10.1030      0.2290    ( 9.6380,10.5679)   ( 9.0209,11.1851) X

X   denotes a row with X values away from the center
```

Figure 8.11 Solution to the ARM problem using the statistical package MINITAB.

to increase by 0.0970 per month. Substituting $t = 48$ into Eq. (39), we obtain the estimate

$$EX_{48} = 5.45 + 0.0970(48) = 10.106$$

for the May 1990 CM1 index value. Since $R^2 = 83.0\%$, the regression equation accounts for 83% of the total variation in our CM1 index data. This gives us a fairly high level of confidence in our estimate of EX_{48}. Of course, the actual

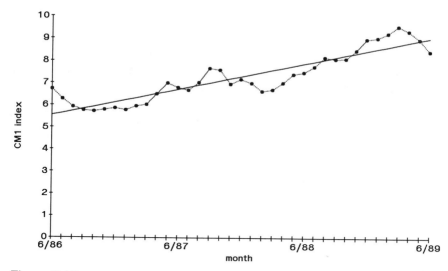

Figure 8.12 Graph of CM1 index versus time showing the regression line for the ARM problem.

value of X_{48} will differ due to random fluctuations. More details about the likely magnitude of these fluctuations will be provided below in the sensitivity analysis.

Finally, step 5. We have concluded that the CM1 index shows a general trend, increasing by about 0.097 points per month. This figure is based on historical observations over the last three years. Projecting on this basis, we obtain the estimate of 10.11 for the May 1990 index figure. This is about 1.1 points higher than the May 1989 index, so in 1990 the author should expect his ARM payments to increase again.

The most important sensitivity analysis question here is the amount of random fluctuation in X_t. We are assuming the linear regression model in Eq. (30), where ε_t is mean zero normal. Our regression package estimated the standard deviation $\sigma \approx 0.4812$ on the basis of the data. In other words, $\varepsilon_t/0.4812$ is approximately standard normal. About 95% of the data points were no more than $\pm 2\sigma$ away from the line in Eq. (41). If this is representative of the magnitude of future fluctuations, then we would expect X_{48} to lie between $10.106 \pm 2\sigma$ with 95% confidence, i.e., we should see

$$9.14 \leq X_{48} \leq 11.07.$$

There is also a more sophisticated method built into the statistics package which takes into account the additional uncertainty involved in estimating EX_{48}. This method yields $9.02 \leq X_{48} \leq 11.19$ at the 95% confidence level. See Fig. 8.11.

Next we consider the sensitivity of our model to unusual data values. We are assuming the linear regression model from Eq. (30). Most of the time the random error ε_t will be small, but there is a small probability that ε_t will be rather large, so that one or more of our data points may lie far off the regression line. We need to consider the sensitivity of our procedure to such anomalous points, which are called *outliers*.

It is not hard to show that the regression line for a set of data points (t_1, y_1), ..., (t_n, y_n) will always pass through the point $(\bar{t}, \bar{y})$ defined by

$$\bar{t} = \frac{t_1 + \cdots + t_n}{n}$$

$$\bar{y} = \frac{y_1 + \cdots + y_n}{n}. \tag{42}$$

In our model we have $\bar{t} = 19$ and $\bar{y} = 7.29$. The regression procedure selects the best-fitting line through the point $(19, 7.29)$. Since the essence of the procedure is to minimize the vertical distance between the regression line and the data points, an outlier will tend to pull the regression line toward itself, regardless of the location of the other data points. The situation gets worse as n gets smaller, because then each individual data point has more influence. It also gets worse as the distance from the base point $(\bar{t}, \bar{y})$ gets larger, since points farther out on the regression line have more leverage.

In Fig. 8.11 the statistical package MINITAB has flagged the data point $(1, 6.73)$ as being unusual. If we plug $t = 1$ into the regression equation, we get

$$\hat{y}_1 = 5.45 + 0.0970(1) = 5.547.$$

The vertical distance or *residual* $y_1 - \hat{y}_1$ is 1.18, which means that this data point is about 2.6 standard deviations above the regression line. In order to ascertain the sensitivity of our model to this outlier, we repeat the regression calculation, leaving out the data point $(1, 6.73)$. Figure 8.13 shows the results of this sensitivity run. The new regression equation is

$$EX_t = 5.30 + 0.103t$$

with $R^2 = 86.2\%$. The predicted value $EX_{48} = 10.24$ and the 95% *prediction interval* is

$$9.24 \leq X_{48} \leq 11.22.$$

This is about the same as before, so we conclude that our model is not too sensitive to this outlier.

The main robustness issue here concerns our choice of a linear model in Eq. (30). More generally, we might assume that

$$X_t = f(t) + \varepsilon_t, \tag{43}$$

```
set c1
2:37
end
set c2
6.27 5.93 5.77 5.72 5.80 5.87 5.78 5.96 6.03 6.50 7.00
6.80 6.68 7.03 7.67 7.59 6.96 7.17 6.99 6.64 6.71 7.01 7.40
7.49 7.75 8.17 8.09 8.11 8.48 8.99 9.05 9.25 9.57 9.36 8.98 8.44
end
regress c2 on 1 predictor c1;
predict 48.
```

```
The regression equation is
C2 = 5.30 + 0.103 C1
```

```
Predictor        Coef        Stdev      t-ratio         p
Constant       5.3045       0.1554       34.13       0.000
C1           0.102634     0.007034       14.59       0.000

s = 0.4385       R-sq = 86.2%       R-sq(adj) = 85.8%
```

```
Analysis of Variance
```

```
SOURCE          DF          SS          MS          F          p
Regression       1       40.924      40.924      212.88      0.000
Error           34        6.536       0.192
Total           35       47.460
```

```
    Fit   Stdev.Fit          95% C.I.              95% P.I.
10.2309      0.2134   ( 9.7972,10.6647)   ( 9.2397,11.2221) X

X  denotes a row with X values away from the center
```

Figure 8.13 Sensitivity analysis for the ARM problem using the statistical package MINITAB.

where $f(t)$ represents the true value of one-year U.S. Treasury bonds at time t, and ε_t represents fluctuations in the market. In this more general setting the linear regression model represents a linear approximation

$$f(t) \approx a + bt, \tag{44}$$

which is good near the base point $(\bar{t}, \bar{y})$. In Figs. 8.11 and 8.13, MINITAB has flagged the point $t = 48$ as being far away from the center point $t = 19$. We have data for $1 \leq t \leq 37$ and strong evidence of a linear relationship over this interval ($R^2 = 83\%$). In other words, the linear approximation in Eq. (44) involves at most a small percent error over this interval. As we move away from this interval, however, it must be expected that the error involved in this linear approximation gets worse.

Our linear regression model is a simple example of a *time series model*. A time series model is a stochastic model of one or more variables that evolve over time. Most economic forecasting is done using time series models. More complex time series models represent the interaction of several variables and dependency in the random fluctuations of these variables. Time series analysis is a branch of statistics. A good place to start learning more about time series models is Box, G. et al., (1976).

8.4 Exercises

1. Reconsider the inventory problem of Example 8.1, but now suppose that the store policy is to order more aquariums if there are less than 2 left in stock at the end of the week. In either case (0 or 1 remaining) the store orders enough to bring the total number of aquariums in stock back up to 3.

    (a) Calculate the probability that the demand for aquariums in a given week exceeds the supply. Use the five-step method, and model as a Markov chain in steady state.

    (b) Perform a sensitivity analysis on the demand rate λ. Calculate the steady-state probability that demand exceeds supply, assuming $\lambda = 0.75, 0.9, 1.0, 1.1,$ and 1.25, and display in graphical form, as in Fig. 8.5.

    (c) Let p denote the steady-state probability that demand exceeds supply. Use the results of part (b) to estimate $S(p, \lambda)$.

2. (Requires a computer algebra system) Reconsider the inventory problem of Example 8.1. In this problem we will explore the sensitivity of the probability p that demand exceeds supply to the demand rate λ.

    (a) Draw the state transition diagram for an arbitrary λ. Show that Eq. (11) is the appropriate state transition probability matrix for this problem.

    (b) Write down the system of equations analogous to Eq. (10) which must be solved to get the steady-state distribution for a general λ. Use a computer algebra system to solve these equations.

    (c) Use the results of part (b) to obtain a formula for p in terms of λ. Graph p versus λ for the range $0 \leq \lambda \leq 2$.

    (d) Use a computer algebra system to differentiate the formula for p obtained in part (c). Calculate the exact sensitivity $S(p, \lambda)$ at $\lambda = 1$.

3. Reconsider the inventory problem of Example 8.1, but now suppose that the

inventory policy depends on recent sales history. Whenever inventory drops to 0, the number of units ordered is equal to 2 plus the number sold over the past week.

(a) Determine the steady-state probability that demand exceeds supply, as well as the average size of a resupply order. Use the five-step method, and model as a Markov chain.

(b) Repeat part (a), but now suppose that weekly demand is Poisson with a mean of 2 customers per week.

4. Reconsider the inventory problem of Example 8.1, but now suppose that 3 additional aquariums are ordered any time that there are less than 2 in stock at the end of the week.

(a) Determine the probability that demand exceeds supply on any given week. Use the five-step method, and model as a Markov chain in steady state.

(b) Use the steady-state probabilities from part (a) to calculate the expected number of aquariums sold per week under this inventory policy.

(c) Repeat part (b) for the inventory policy in Example 8.1.

(d) Suppose that the store makes a profit of $5 per 20-gallon aquarium sold. How much would the store gain by implementing the new inventory policy?

5. For the purposes of this problem we will consider the stock market to be in one of three states:

$$1\text{—Bear market}$$
$$2\text{—Strong bull market}$$
$$3\text{—Weak bull market}$$

Historically, a certain mutual fund gained -3%, 28%, and 10% annually when the market was in states 1, 2, and 3 respectively. Assume that the state transition probability matrix

$$P = \begin{pmatrix} 0.90 & 0.02 & 0.08 \\ 0.05 & 0.85 & 0.10 \\ 0.05 & 0.05 & 0.90 \end{pmatrix}$$

applies to the weekly change of state in the stock market.

(a) Determine the steady-state distribution of market state.

(b) Suppose that $10,000 is invested in this fund for 10 years. Determine the expected total yield. Does the order of state transitions make any difference?

(c) In the worst-case scenario the long-run expected proportion of time in each market state is 40%, 20%, and 40%, respectively. What is the effect on the answer to part (b)?

(d) In the best-case scenario the long-run expected proportion of time in each market state is 10%, 70%, and 20%, respectively. What is the effect on the answer to part (b)?

(e) Does this mutual fund offer a better investment opportunity than a money market fund currently yielding about 8%? Consider that the money market fund offers a lower risk.

6. This exercise shows the equivalence of the two formulations of a Markov process model.

(a) Suppose that $T_{i1}, \ldots, T_{im}$ are independent random variables and that T_{ij} is negative exponential with rate parameter $a_{ij} = p_{ij}\lambda_i$. Assume that $\sum p_{ij} = 1$ and $\lambda_i > 0$. Show that

$$T_i = \min(T_{i1}, \ldots, T_{im})$$

has a negative exponential distribution with rate parameter λ_i. [Hint: Use the fact that

$$\Pr\{T_i > x\} = \Pr\{T_{i1} > x, \ldots, T_{im} > x\}$$

for all $x > 0$.]

(b) Suppose that $m = 2$ so that $T_i = \min(T_{i1}, T_{i2})$ and show that $\Pr\{T_i = T_{i1}\} = p_{i1}$. [Hint: Use the fact that

$$\Pr\{T_i = T_{i1}\} = \Pr\{T_{i2} > T_{i1}\}$$
$$= \int_0^\infty \Pr\{T_{i2} > x\} f_{i1}(x)\,dx,$$

where $f_{i1}(x)$ is the probability density function of the random variable T_{i1}.]

(c) Use the results of (a) and (b) to show that in general $\Pr\{T_i = T_{ij}\} = p_{ij}$.

7. Reconsider the forklift problem of Example 8.3, but now suppose that a reserve fleet of m additional trucks has been purchased.

(a) How large should m be in order to guarantee that 95% of the time there will be at least 27 working forklift trucks available?

(b) Suppose that a preventative maintenance policy is available which has the effect of prolonging the average time between repairs for a typical forklift truck to about nine months. Answer the same question as in part (a).

(c) The cost of a new forklift truck, amortized over the expected number of years of service, is about $2,800 per year. The cost of the preventative maintenance program is $600 per truck per year. Which of the two plans, (a) or (b), is the most cost-effective way of ensuring an adequate fleet of operating vehicles?

(d) There is some uncertainty as to the actual improvement in the time between repairs as a result of preventative maintenance. What is the minimum improvement for which plan (b) beats plan (a)?

8. Five locations are connected by radio. The radio link is active 20% of the time, and there is no radio activity the remaining 80% of the time. The main location sends radio messages with an average duration of 30 seconds, and the remaining four locations send messages that average 10 seconds in length. Half of all radio messages originate at the main location, with the remaining proportion equally divided among the other four locations.

(a) For each location determine the steady-state probability that this location is sending a message at any given time. Use the five-step method, and model as a Markov process.

(b) A monitoring station samples from the radio emissions on this network once every five minutes. How long on average does it take until the monitor finds a message in progress from one particular location?

(c) The monitor can identify the source of a radio message if the message lasts at least three seconds after the monitor begins listening. How long does it take until the monitor locates one particular location?

(d) Perform a sensitivity analysis to see how the results of part (c) are affected by the percent utilization of the radio frequency (currently 20%).

9. A gasoline service station has two pumps, each of which can service two cars at a time. If all pumps are busy, cars will queue up in a single line to await a free pump. Since the station operates in a competitive environment, it can be expected that customers encountering long lines at this station will take their business elsewhere.

(a) Construct a model that can be used to predict both the steady-state probability of a waiting line and the expected length of the line. Use the five-step method, and model as a Markov process. You will have to make some additional assumptions about customer demand, service time, and balking (refusing to join a queue).

(b) Use the model of part (a) to estimate the fraction of potential business lost because of customer balking. Consider a range of possible levels of customer demand.

(c) What is the easiest way to infer the level of customer demand (i.e., potential sales) from data to which the station manager has access?

(d) Under what circumstances would you recommend that the station purchase additional pumps?

Table II. Per capita income versus population density for the 10 poorest Asian nations.

Country	Per capita income (in 1982 dollars)	Population density (pop. per sq. mi.)
Nepal	168	290
Kampuchea	117	101
Bangladesh	122	1740
Burma	171	139
Afghanistan	172	71
Bhutan	142	76
Vietnam	188	458
China	267	284
India	252	578
Laos	325	45

10. A certain form of one-celled creature reproduces by cell division, producing two offspring. The mean lifetime before cell division is one hour, and each individual cell has a 10% chance of dying before it can reproduce.
    (a) Construct a model that represents the evolution of population size over time. Use a Markov process, and draw a rate diagram.
    (b) Describe in general terms what you would expect to happen to the population level over time.
    (c) What is the problem with applying steady-state results to this model?
11. Table II gives per capita income in 1982 dollars and population density in population per square miles for the 10 poorest counties in Asia (source: Webster's New World Atlas, (1988)).
    (a) Does the data support the proposition that prosperity is linked to population density? Use linear regression to obtain a formula that predicts per capita income as a linear function of population density.
    (b) What percentage of the total variation in per capita income can be attributed to variations in population density?
    (c) What is the effect on your answers to parts (a) and (b) if we leave out of our analysis the country (Bangladesh) with population density of more than 1,000 per square mile?
    (d) On the basis of your regression model, estimate the potential benefit for the citizens of a poor Asian country that manages to reduce its population by 25%.

12. Reconsider problem 11(a) above, but now obtain the regression line by solving the underlying optimization problem by hand. Letting t_i denote the population density and y_i the per capita income of country i, the goodness-of-fit for a candidate regression line

$$y = a + bt$$

    is given by Eq. (33) in the text. Plug in the data points (t_i, y_i) and compute the function $F(a, b)$. Then obtain the best-fitting line by minimizing $F(a, b)$ over the set of all $(a, b) \in \mathbb{R}^2$.

13. Reconsider the ARM problem of Example 8.5. Suppose that only the data for the period June 1986 to June 1988 were known, and we were attempting to predict the May 1989 value of the CM1 index.
    (a) Use a computer or calculator implementation of linear regression to obtain the regression line for this data.
    (b) What is the predicted value of the May 1989 CM1 index according to the regression model of part (a)?
    (c) What is the value of R^2 for the model in part (a)? How would you interpret this value?
    (d) Compare to the actual value of the May 1989 CM1 index. How close was the predicted value? Was it within two standard deviations?

14. Repeat Exercise 13 above, but use the numbers for the TB3 index.

15. Reconsider the ARM problem of Example 8.5. Use a computer program for multiple regression to predict future trends in the CM1 index by fitting a second-degree polynomial to the data. After you input the time index $t = 1, 2, 3, \ldots$ in one column and the CM1 data to another column, prepare another column of data containing the second power of the time index numbers $t^2 = 1, 4, 9, \ldots$. Multiple regression will give the best-fitting second-degree polynomial

$$CM1 = a + bt + ct^2,$$

    and R^2 can be interpreted as before. This technique is known as *polynomial least squares*.
    (a) Use a computer implementation of multiple linear regression to obtain a formula that predicts the CM1 index as a quadratic function of t.
    (b) Use the formula obtained in part (a) to predict the expected May 1990 value for the CM1 index.
    (c) What percent of the total variation in the CM1 index can be accounted for by using this model?
    (d) Compare the R^2 value for this multiple regression model with what was done in Section 8.3. Which model gives the best fit?

Table III.   Response-time data for the facility location problem.

Distance (miles)	Drive time (minutes)
1.22	2.62
3.48	8.35
5.10	6.44
3.39	3.51
4.13	6.52
1.75	2.46
2.95	5.02
1.30	1.73
0.76	1.14
2.52	4.56
1.66	2.90
1.84	3.19
3.19	4.26
4.11	7.00
3.09	5.49
4.96	7.64
1.64	3.09
3.23	3.88
3.07	5.49
4.26	6.82
4.40	5.53
2.42	4.30
2.96	3.55

16. (Response-time formula from Example 3.2) A suburban community intends to replace its old fire station with a new facility. As part of the planning process, response-time data were collected for the past quarter. It took an average of 3.2 minutes to dispatch the fire crew. The dispatch time was found to vary only slightly. The time for the crew to reach the scene of the fire (drive time) was found to vary significantly depending on the distance to the scene. The data on drive time are displayed in Table III.

(a) Use linear regression to obtain a formula that predicts drive time as a linear function of distance traveled. Then determine a formula for total response time, including dispatch time.

(b) What percentage of the total variation in drive time is accounted for by the formula you came up with in part (a)?

(c) Draw a graph of drive time versus distance for the data in Table III. Do the data seem to indicate a linear trend?

(d) Plot the regression line from part (a) on the graph made in part (c). Does the line seem to be a good predictor for this data?

17. (Continuation of Exercise 16) Another way to get a formula that relates drive time d to distance r is to use an exponential model. Suppose that the underlying relationship between d and r is of the form $d = ar^b$. Taking logarithms of both sides yields the relationship

$$\ln d = \ln a + b \ln r.$$

Then linear regression can be used to estimate the parameters $\ln a$ and b in this linear equation.

(a) Transform the data in Table III by taking logarithms of both the drive time d and the distance r. Plot $\ln d$ versus $\ln r$. Does your graph suggest a linear relationship between $\ln d$ and $\ln r$?

(b) Use linear regression to obtain a formula that predicts $\ln d$ as a linear function of $\ln r$. Then determine a formula for total response time, including dispatch time, as a function of the distance d to the fire. Compare to the formula in Example 3.2.

(c) What is the value of R^2 for your regression model in part (b)? How do you interpret this number?

(d) Plot drive time d versus distance r, and then sketch a graph of the formula for d as a function of r, which you determined in part (b). Does the exponential model seem to give a good fit to this data?

(e) Compare the results of parts (c) and (d) above with the results of Exercise 16. Which model seems to give a better fit to this data? Justify your answer.

Further Reading

Arrow, K. et al. (1958). *Studies in the Mathematical Theory of Inventory and Production*, Stanford University Press, Stanford, California.

Billingsley, P. (1979). *Probability and Measure*, Wiley, New York.

Box, G. and Jenkins, G. (1976). *Time Series Analysis, Forecasting, and Control*, Holden-Day, San Francisco.

Çinlar, E. (1975). *Introduction to Stochastic Processes*, Prentice–Hall, Englewood Cliffs, New Jersey.

Cornell, R., Flora, J. and Roi, L. *The Statistical Evaluation of Burn Care*, UMAP module 553.

Freedman, D. et al. (1978). *Statistics*, W. W. Norton, New York.

Giordano, F., Wells, M. and Wilde, C. *Dimensional Analysis*, UMAP module 526.

Giordano, M., Jaye, M. and Weir, M. *The Use of Dimensional Analysis in Mathematical Modeling*, UMAP module 632.

Hillier, F. and Lieberman, G. (1990). *Introduction to Operations Research*, 5th Ed., Holden-Day, Oakland CA.

Hogg, R. and Tanis, E. (1988). *Probability and Statistical Inference*, Macmillan, New York.

Huff, D. (1954). *How to Lie with Statistics*, W. W. Norton, New York.

Kayne, H. *Testing a Hypothesis: t–Test for Independent Samples*, UMAP module 268.

Keller, M. *Markov Chains and Applications of Matrix Methods: Fixed-Point and Absorbing Markov Chains*, UMAP modules 107 and 111.

Knapp, T. *Regression Toward the Mean*, UMAP module 406.

Meerschaert, M. and Cherry, W. P. (1988). Modeling the behavior of a scanning radio communications sensor, *Naval Research Logistics Quarterly*, Vol. 35, 307–315.

Travers, K., and Heeler, P. *An Iterative Approach to Linear Regression*, UMAP module 429.

Yates, F. *Evaluating and Analyzing Probabilistic Forecasts*, UMAP module 572.

Chapter Nine

Simulation of Probability Models

Computational methods for optimization are important, because most optimization problems are too difficult to be solved analytically. For dynamic models, it is often possible to determine steady-state behavior analytically, but the study of transient (time-dependent) behavior requires computer simulation. Probability models are even more complex. Models with no time dynamics can sometimes be solved analytically, and steady-state results are available for the simplest stochastic models. But for the most part, probability models are solved by simulation. In this chapter we will discuss some of the most generally useful simulation methods for probability models.

9.1 Monte Carlo Simulation

Questions involving the transient or time-dependent behavior of stochastic models are difficult to resolve analytically. Monte Carlo simulation is a general modeling technique that is usually effective for such problems. The construction of Monte Carlo simulation software can be time-consuming, and the repeated simulation runs needed for accuracy and sensitivity analysis can become prohibitively expensive. Even so, Monte Carlo simulation models continue to enjoy a very wide appeal. They are easy to conceptualize, easy to explain, and they are the only viable method for the modeling of many complex stochastic systems. A Monte

Variables: $X_t = \begin{cases} 0 & \text{if no rain on day } t \\ 1 & \text{if rain on day } t \end{cases}$

Assumptions: $X_1, X_2, \ldots, X_7$ are independent
$\Pr\{X_t = 0\} = \Pr\{X_t = 1\} = 1/2$

Objective: Determine the probability that
$X_t = X_{t+1} = X_{t+2} = 1$ for some
$t = 1, 2, 3, 4,$ or 5

Figure 9.1 Results of step 1 of the rainy day problem.

Carlo simulation models random behavior. It can be based on any simple ran-
domizing device, such as coin flips or the roll of dice, but typically a computer
pseudorandom number generator is used. Because of the random element, each
repetition of the model will produce different results.

Example 9.1   Arriving on your vacation you are dismayed to learn that the local
weather service forecasts a 50% chance of rain every day this week. What are the
chances of 3 consecutive rainy days?

We will use the five-step method. Step 1 is to formulate a question. In the
process we assign variable names to quantities of interest, and we clarify our
assumptions about these variables. Then we state a question in terms of these
variables. See Figure 9.1 for the results of step 1.

Step 2 is to select the modeling approach. We will use Monte Carlo simulation.

> *Monte Carlo simulation* is a technique that can be applied to any probability
> model. A probability model includes a number of random variables and must
> also specify the probability distribution for each of these random variables.
> Monte Carlo simulation uses a randomizing device to assign a value to each
> random variable, in accordance with its probability distribution. Since the
> results of the simulation depend on random factors, subsequent repetitions
> of the same simulation will reproduce different results. Usually a Monte
> Carlo simulation will be repeated a number of times in order to determine an
> average or expected outcome.

> Monte Carlo simulation is typically used to estimate one or more measures
> of system performance (MOPs). Repeated simulations can be considered
> as independent random trials. For the moment let us consider the situation
> where there is only one simulation parameter Y to be examined. Repeated
> simulation produces the results $Y_1, Y_2, \ldots, Y_n$, which we may consider as
> independent and identically distributed random variables, whose distribution

is unknown. By the strong law of large numbers we know that

$$\frac{Y_1 + \cdots + Y_n}{n} \to EY \tag{1}$$

as $n \to \infty$. Hence, we should use the average of $Y_1, \ldots, Y_n$ to estimate the true expected value of Y. We also know that, letting

$$S_n = Y_1 + \cdots + Y_n,$$

the central limit theorem implies that

$$(S_n - n\mu)/\sigma\sqrt{n}$$

is approximately standard normal for large n, where $\mu = EY$ and $\sigma^2 = VY$. For most cases the normal approximation is fairly good when $n \geq 10$. Even though we do not know μ or σ, the central limit theorem still gives some important insight. The difference between the observed average S_n/n and the true mean $\mu = EY$ is

$$\frac{S_n}{n} - \mu = \frac{\sigma}{\sqrt{n}}\left(\frac{S_n - n\mu}{\sigma\sqrt{n}}\right), \tag{2}$$

so we can expect the variation in the observed mean to tend to zero about as fast as $1/\sqrt{n}$. In other words, to get one more decimal place of accuracy in EY would require 100 times as many repetitions of the simulation. More sophisticated statistical analysis is possible, but the basic idea is now very clear. We will have to be satisfied with fairly rough approximations of average behavior if we are to use Monte Carlo simulation.

As a practical matter, there are many sources of error and variation in a modeling problem, and the additional variation produced by Monte Carlo simulation is not typically the most serious of these. A judicious application of sensitivity analysis will suffice to ensure that the results of a simulation are used properly.

Moving on to step 3, we now need to formulate the model. Figure 9.2 gives an algorithm for Monte Carlo simulation of our vacation problem. As in Chapter 3, the notation Random $\{S\}$ denotes a point selected at random from the set S. In our simulation, each day's weather is represented by a random number from the interval $[0, 1]$. If the number turns out to be less than p, we assume that this is a rainy day. Otherwise it is a sunny day. Then p is the probability that any one day is rainy. The variable C simply counts the number of consecutive rainy days. Figure 9.3 shows a slightly modified algorithm. The modified version repeats the Monte Carlo simulation n times and counts the number of rainy weeks (i.e., the number of weeks in which it rains 3 days in a row). The notation

$$Y \leftarrow \text{Rainy Day Simulation } (p)$$

Algorithm: Rainy Day Simulation

Variables: p = probability of one rainy day

$$X(t) = \begin{cases} 1 & \text{if rain on day } t \\ 0 & \text{otherwise} \end{cases}$$

$$Y = \begin{cases} 1 & \text{if } \geq 3 \text{ consecutive rainy days} \\ 0 & \text{otherwise} \end{cases}$$

Input: p

Process: Begin
    $Y \leftarrow 0$
    $C \leftarrow 0$
    for $t = 1$ to 7 do
        Begin
        if Random $\{[0, 1]\} < p$ then
            $X(t) = 1$
        else
            $X(t) = 0$
        if $X(t) = 1$ then
            $C \leftarrow C + 1$
        else
            $C \leftarrow 0$
        if $C \geq 3$ then $Y \leftarrow 1$
        End
    End

Output: Y

Figure 9.2 Pseudocode for Monte Carlo simulation of the rainy day problem.

indicates that we evaluate the output variable Y by running the rainy day simulation from Fig. 9.2 with input variable p.

Step 4 is to solve the problem. We ran a computer implementation of the algorithm in Fig. 9.3 with $p = 0.5$ and $n = 100$. The simulation counted 43 rainy weeks out of 100. On this basis we would estimate a 43% chance of a rainy week. Several additional runs were made to confirm these results. In every case the simulation counted around 40 rainy weeks out of 100. Given the likely magnitude of error in the 50% estimated chance of rain, this is about as much accuracy as we will need for this problem. More details on the sensitivity of our simulation results

Algorithm:	Repeated Rainy Day Simulation

Variables:
p = probability of 1 rainy day
n = number of weeks to simulate
S = number of rainy weeks

Input: p, n

Process:
Begin
$\quad S \leftarrow 0$
$\quad$for $k = 1$ to n do
$\qquad$Begin
$\qquad\quad Y \leftarrow$ Rainy Day Simulation (p)
$\qquad\quad S \leftarrow S + Y$
$\qquad$End

Output: S

Figure 9.3 Pseudocode for repeated Monte Carlo simulation to determine average behavior in the rainy day problem.

to random factors will be discussed later, in the section on sensitivity analysis.

Finally, step 5. Arriving on your vacation, you find that the local weather service predicts a 50% chance of rain every day for a week. A simulation indicates that, if this forecast is correct, there is a 40% chance that there will be at least 3 consecutive rainy days this week. These results apply to sunshine as well as rain, and so, to end on a somewhat more optimistic note, let us point out that there is a 50% chance of sunshine every day this week, and a 40% chance of at least 3 consecutive days of sunshine. Enjoy your vacation!

We will begin our sensitivity analysis by examining the sensitivity of our simulation results to random factors. Each model run simulates $n = 100$ weeks of vacation and counts the number of rainy weeks. In the terminology of step 2, our MOP is Y, where $Y = 1$ indicates a rainy week, and $Y = 0$ indicates otherwise. Our model simulates $n = 100$ independent random variables $Y_1, \ldots, Y_n$, all of which have the same distribution as Y. Here $Y_k = 1$ indicates that week k was a rainy week. Our model outputs the random variable $S_n = Y_1 + \cdots + Y_n$, which represents the number of rainy weeks. Let

$$q = \Pr\{Y = 1\} \qquad (3)$$

denote the probability of a rainy week. It is not hard to calculate that

$$\mu = EY = q$$
$$\sigma^2 = VY = q(1 - q). \tag{4}$$

Our first model run output $S_n = 43$. On this basis, we can use the strong law of large numbers, Eq. (1), to estimate that

$$q = EY \approx S_n/n = 0.43. \tag{5}$$

How good is this estimate? By the central limit theorem, we obtain from Eq. (2) that S_n/n is unlikely to differ from $\mu = q$ by more than $2\sigma/\sqrt{n}$, since a standard normal random variable is 95% certain to have absolute value less than 2. Using Eqs. (4) and (5), we would conclude that our estimate in Eq. (5) is within

$$2\sqrt{(0.43)(0.57)/100} \approx 0.1 \tag{6}$$

of the true value of q.

A more elementary way to investigate the sensitivity of our simulation results to random factors is to compare the results of a number of model runs. Figure 9.4 shows the results of 40 model runs, each of which simulates 100 weeks of vacation. All of these model runs lead to estimates $S_n/n \approx 0.4$, and none is

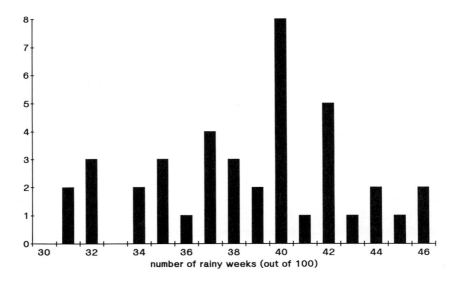

Figure 9.4 Histogram showing the distribution of the number of rainy weeks out of 100 for the rainy day problem.

outside of the interval 0.4 ± 0.1. It also appears that the distribution of S_n is approximately normal.

We should also examine the sensitivity of our simulation results to the forecast 50% chance of rain. Figure 9.5 shows the results of 10 additional model runs for each of the cases $p = .3, .4, .5, .6$, and $.7$, where p is the probability of one rainy day. The boxes connected by the dotted line show the average outcome (fraction of rainy weeks), and the vertical bars show the range of outcomes for each case. For a 40% chance of rain each day the probability of at least 3 consecutive rainy days in a week is around 20%, and so forth. While the probability of 3 straight days of rain varies quite a bit, it seems safe to say that if the chance of rain each day is moderate, then so is the chance of rain 3 straight days in one week.

What about robustness? We should examine the critical assumptions that make up the structure of the model. In step 1 we assumed that the indicator variables $X_1, \ldots, X_7$ were independent and identically distributed. In other words, the chance of rain is the same each day, and the weather on one day is independent of the weather on any other day. Suppose instead that $\Pr\{X_t = 1\}$ varies with t, still keeping the independence assumption. We can use our sensitivity analysis results to obtain upper and lower bounds on the probability of 3 consecutive days of rain, reasoning that this probability should be monotone increasing as any $\Pr\{X_t = 1\}$ increases. A higher chance of rain on day t means a higher probability of 3

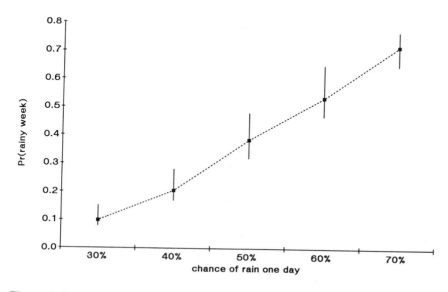

Figure 9.5 Graph of the probability of a rainy week versus the chance of rain one day for the rainy day problem.

straight days of rain, all else being equal. Then if all $\Pr\{X_t = 1\}$ are between 0.4 and 0.6 (a 40% to 60% chance of rain), the probability of 3 days of rain is between .2 and .6. Our model is quite robust in this regard, partly due to the fact that we are not looking for a lot of accuracy in our answer.

Suppose now that $\{X_t\}$ are not independent. For example, we could model $\{X_t\}$ as a Markov chain on the state space $\{0, 1\}$. This implies that the chance of rain today depends on the weather yesterday. Local weather forecasts rarely contain the kind of information necessary to formulate such a model, particularly the state transition probabilities. However, we could always guess at these and then use sensitivity analysis to ensure that we have not assumed too much. The kind of question we are asking here cannot be answered using a steady-state analysis, because it concerns the time-dependent or transient behavior of the stochastic process. Such questions are usually difficult to handle analytically, even by advanced techniques. This is one reason why Monte Carlo simulation is so widely used.

9.2   The Markov Property

A stochastic process is said to have the Markov property if the information contained in the current state of the process is all that is needed to determine the probability distribution of future states. Markov chains and Markov processes were introduced in Chapter 8. Both have the Markov property. Monte Carlo simulation of a stochastic process is much simpler in the presence of the Markov property, because it reduces the quantity of information that needs to be stored in the computer.

Example 9.2   We reconsider the docking problem of Example 4.3, now taking into account the random element. Our basic assumptions are summarized in Fig. 4.7. Our goal or objective, as before, is to determine the success of our control procedure to match velocities.

We will use the five-step method. Our starting point is Fig. 4.7, but we must also make some assumptions about the variables a_n, c_n, and w_n. Ideally we would perform some experiments and collect data on such factors as astronaut response time and the time it takes to read or manipulate controls. In the absence of such data we would attempt to make some reasonable assumptions consistent with what is known about similar situations.

The random variable that represents the most uncertainty (i.e., the greatest variance) is c_n, the time it takes to make a control adjustment. This variable represents the time it takes to observe the rate of closing, calculate the desired acceleration adjustment, and carry out the adjustment. We will assume that it takes roughly 1 second to observe the rate of closing, 2 seconds to calculate the adjustment, and 2 seconds to make the adjustment. The actual time to carry out each phase is random. Let R_n denote the time to read the closing velocity, S_n the time to calculate the desired adjustment, and T_n the time to make the adjustment. Then $ER_n = 1$ second, and $ES_n = ET_n = 2$ seconds. We need to make a reasonable assumption about the distributions of these random variables. It seems reasonable to suppose that they are all nonnegative, mutually independent, and that outcomes close to the mean are most likely. There is a wide variety (infinite!) of distributions that fit this general description, and as of now we have no particular reason to prefer one over another. This being the case, we will refrain from specifying the exact distribution at this time. One of our random variables is $c_n = R_n + S_n + T_n$. The others are w_n, the waiting time before the next control adjustment, and a_n, the acceleration after this control adjustment. We will assume that $a_n = -kv_n + \varepsilon_n$, where ε_n is a (small) random error. We assume that ε_n is normally distributed with mean zero. The variance of ε_n depends on both the skill of the human operator and on the sensitivity of the control mechanism. We will assume that we can typically achieve an accuracy of about ± 0.05 m/sec^2, and so we set the standard deviation of ε_n at $\sigma = 0.05$. The waiting time w_n will depend on c_n if we are trying to maintain a fixed time between control adjustments of 15 seconds total. We will assume that $w_n = 15 - c_n + E_n$, where E_n is a small random error. Assume E_n has mean zero, is normal, and that astronaut response-time limitations imply a standard deviation of 0.1 seconds.

We should now consider the analysis objectives. We want to determine the success of our control procedure. Certainly we are interested in seeing that $v_n \to 0$. Simulation can determine that. But we also have the opportunity to gather information on other aspects of system performance. We should decide at this time, as a part of step 1, what measures of performance (MOPs) we want to track. The selected MOPs should represent important quantitative information that can be used as the basis for comparison with computing control procedure options. Assume that our initial closing velocity is 50 m/sec, and that the velocity-matching process is considered successful when closing velocity has been reduced to 0.1 m/sec. We would be most interested in the total time it takes to succeed. This will be our measure of performance. At this point, to conclude step 1, we summarize in Figure 9.6.

Step 2 is to select the modeling approach. We will use Monte Carlo simulation based on the Markov property.

> The general idea is as follows. At each time step n there is a vector X_n that describes the current state of the system. The sequence of random vectors $\{X_n\}$ is assumed to have the Markov property. In other words, the current

Variables: t_n = time of nth velocity observation (sec)
                   v_n = velocity at time t_n (m/sec)
                   c_n = time to make nth control adjustment (sec)
                   a_n = acceleration after nth control adjustment (m/sec^2)
                   w_n = wait before $(n+1)$th observation (sec)
                   R_n = time to read velocity (sec)
                   S_n = time to calculate adjustment (sec)
                   T_n = time to make adjustment (sec)
                   ε_n = random error incontrol adjustment (m/sec^2)
                   E_n = random error in waiting time (sec)

Assumptions: $t_{n+1} = t_n + c_n + w_n$
                   $v_{n+1} = v_n + a_{n-1}c_n + a_n w_n$
                   $a_n = -kv_n + \varepsilon_n$
                   $c_n = R_n + S_n + T_n$
                   $w_n = 15 - c_n + E_n$
                   $v_0 = 50, \ t_0 = 0$
                   $E R_n = 1, \ E S_n = E T_n = 2$, and the
                   distribution of R_n, S_n, T_n is yet
                   to be specified.
                   ε_n is normal mean 0, standard
                   deviation 0.05.
                   E_n is normal mean 0, standard
                   deviation 0.1.

Objective: Determine $T = \min\{t_n : |v_n| \leq 0.1\}$

Figure 9.6 Results of step 1 of the docking problem with random factors.

state X_n contains all of the information needed to determine the probability distribution of the next state, X_{n+1}.

The general structure of the simulation is as follows. First we initialize variables and read data files. At this stage we must specify the initial state X_0. Next we enter a loop that repeats until an end condition is satisfied. In the loop we use X_n to specify the distribution of X_{n+1}, and then we use a random number generator to determine X_{n+1} according to that distribution. We must also calculate and store any information needed to generate the simulation

```
Begin
Read data
Initialize X₀
While (not done) do
        Begin
        Determine distribution of Xₙ₊₁ using Xₙ
        Use Monte Carlo method to determine Xₙ₊₁
        Update records for MOPs
        End
Calculate and output MOPs
End
```

Figure 9.7 Algorithm for the general Markovian simulation.

MOPs. Once the end condition occurs, we exit the loop and output the MOPs. Then we are done. The algorithm for this simulation is illustrated by Figure 9.7.

We need to discuss the inner loop in some detail. Suppose for now that the state vector X_n is one-dimensional. Let

$$F_\Theta(t) = \Pr\{X_{n+1} \le t \mid X_n = \Theta\}.$$

The value of $\Theta = X_n$ determines the probability distribution of X_{n+1}. The function F_Θ maps the state space

$$E \subseteq \mathbb{R}$$

onto the interval [0, 1]. There are widely available methods for generating a random number in [0, 1], and these can be used to generate a random variable with distribution F_Θ. Since

$$y = F_\Theta(x)$$

maps

$$E \to [0, 1],$$

the inverse function

$$x = F_\Theta^{-1}(y)$$

maps

$$[0, 1] \to E.$$

If U is a random variable uniformly distributed over [0, 1] (i.e., the density function of U is the function equal to 1 on [0, 1] and equal to 0 elsewhere), then $X_{n+1} = F_\Theta^{-1}(U)$ has distribution F_Θ, since given $X_n = \Theta$,

$$\begin{aligned}
\Pr\{X_{n+1} \le t\} &= \Pr\{F_\Theta^{-1}(U) \le t\} \\
&= \Pr\{U \le F_\Theta(t)\} \\
&= F_\Theta(t)
\end{aligned} \tag{7}$$

because
$$\Pr\{U \leq x\} = x \text{ for } 0 \leq x \leq 1.$$

Example 9.3   Let $\{X_n\}$ denote a stochastic process where X_{n+1} has a negative exponential distribution with rate parameter X_n. Determine the first passage time

$$T = \min\{n : X_1 + \cdots + X_n \geq 100\},$$

assuming that $X_0 = 1$.

We will present a computer simulation to solve this problem once we discuss the details of generating X_{n+1} from X_n. Letting $\Theta = X_n$, the density function of X_{n+1} is
$$f_\Theta(x) = \Theta e^{-\Theta x}$$
on $x \geq 0$. The distribution function is
$$F_\Theta(x) = 1 - e^{-\Theta x}.$$

Setting
$$y = F_\Theta(x) = 1 - e^{-\Theta x}$$

and inverting, we obtain
$$x = F_\Theta^{-1}(y) = -\ln(1 - y)/\Theta.$$

Hence we can let
$$X_{n+1} = -\ln(1 - U)/\Theta,$$

where U is a random number between 0 and 1.   See Figure 9.8 for the complete simulation algorithm.

The above discussion provides a method of generating random variables with any prescribed distribution. While useful in theory, in practice there is sometimes a catch. For many distributions, such as the normal distribution, it is not easy to compute the inverse function F_Θ^{-1}. We can always circumvent this difficulty by interpolating from a table of functional values, but in the case of a normal distribution there is an easier way.

The central limit theorem guarantees that for any sequence of independent, identically distributed random variables $\{X_n\}$ with mean μ and variance σ^2, the normalized partial sums

$$((X_1 + \cdots + X_n) - n\mu)/\sigma \sqrt{n}$$

tend to a standard normal distribution. Suppose $\{X_n\}$ are uniform on $[0, 1]$. Then

$$\mu = \int_0^1 x \cdot dx = 1/2$$

$$\sigma^2 = \int_0^1 (x - 1/2)^2 \, dx = 1/12,$$

(8)

Algorithm:	First Passage Time Simulation (Example 9.3)
Variables:	X = initial state variable N = first passage time
Input:	X

Process:	Begin
	$S \leftarrow 0$
	$N \leftarrow 0$
	until $(S \geq 100)$ do
	    Begin
	    $U \leftarrow$ Random $\{[0, 1]\}$
	    $R \leftarrow X$
	    $X \leftarrow -\ln(1 - U)/R$
	    $S \leftarrow S + X$
	    $N \leftarrow N + 1$
	    End
	End

Output:	N

Figure 9.8 Pseudocode for the Markovian simulation for Example 9.3.

and so for n sufficiently large, the random variable

$$Z = \frac{(x_1 + \cdots + X_n) - n/2}{\sqrt{n/12}} \tag{9}$$

is approximately standard normal. For most purposes a value of $n \geq 10$ is sufficient. We will use $n = 12$ to eliminate the denominator in Eq. (10). Given a standard normal random variable Z, another normal random variable Y with mean μ and standard deviation σ can be obtained by setting

$$Y = \mu + \sigma Z. \tag{10}$$

Figure 9.9 shows a simple algorithm for generating normal random variables with a specified mean and variance.

Returning to the docking problem, we begin step 3. Our first concern is to identify our state variables. In this case we can take

$$
\begin{aligned}
T &= t_n \\
V &= v_n \\
A &= a_n \\
B &= a_{n-1}
\end{aligned} \tag{11}
$$

Algorithm: Normal Random Variable

Variables: μ = mean
                        σ = standard deviation
                        Y = normal random variable with mean μ, standard deviation σ

Input: μ, σ

Process: Begin
                        $S \leftarrow 0$
                        for $n = 1$ to 12 do
                            Begin
                                $S \leftarrow S +$ Random $\{[0, \ 1]\}$
                            End
                        $Z \leftarrow S - 6$
                        $Y \leftarrow \mu + \sigma Z$
                        End

Output: Y

Figure 9.9 Pseudocode for Monte Carlo simulation of a normal random variable.

as our state variables. Since our only MOP is already a state variable, we will not need to initialize or update any additional variables for that purpose. Figure 9.10 gives the algorithm for our docking simulation. The notation Normal (μ, σ) denotes the output of the normal random variable algorithm described in Fig. 9.9 above. For now we will assume that

$$c_n = R_n + S_n + T_n$$

has a normal distribution with a mean of $\mu = 5$ seconds and a standard deviation of $\sigma = 1$ second.

Figure 9.11 shows the results of 20 simulation runs. In these runs the docking time ranged between 156 and 604 seconds, with an average docking time of 305 seconds.

We have constructed a Monte Carlo simulation of a velocity-matching exercise for spacecraft manual docking. Our simulation takes into account the random factors inherent in this man–machine system. Based on what we believe to be a reasonable set of assumptions on operator performance, the model indicates a

Algorithm:	Docking Simulation
Variables:	k = control parameter
	n = number of control adjustments
	$T(n)$ = time (sec)
	$V(n)$ = current velocity (ft/sec)
	$A(n)$ = current acceleration (ft/sec^2)
	$B(n)$ = previous acceleration (ft/sec^2)
Input:	$T(0)$, $V(0)$, $A(0)$, $B(0)$, k
Process:	Begin
	$n \leftarrow 0$
	while $|V(n)| > 0.1$ do
	Begin
	$c \leftarrow$ Normal $(5, 1)$
	$B(n) \leftarrow A(n)$
	$A(n) \leftarrow$ Normal $(-k\,V(n), 0.05)$
	$w \leftarrow$ Normal $(15 - c, 0.1)$
	$T(n) \leftarrow T(n) + c + w$
	$V(n) \leftarrow V(n) + c\,B(n) + w\,A(n)$
	$n \leftarrow n + 1$
	End
	End
Output:	$T(n)$

Figure 9.10 Pseudocode for Monte Carlo simulation of the docking problem.

wide variance in the time to complete the docking procedure. For example, the time to match velocities starting from a relative velocity of 50 m/sec, and using a 1:50 control factor, averaged about 5 minutes. But outcomes of less than 3 or greater than 7 minutes are not uncommon. The major source of variation is the time it takes the pilot to complete the control adjustment procedure.

One important parameter for sensitivity analysis is the standard deviation of c_n, the time to make a control adjustment. We have assumed that the standard deviation of c_n is $\sigma = 1$. Figure 9.12 shows the results of 20 additional model

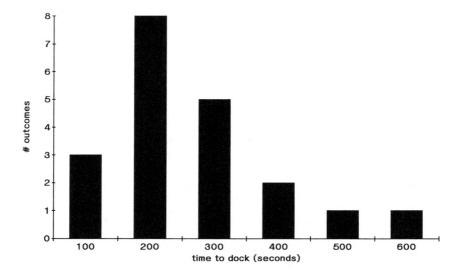

Figure 9.11 Histogram showing the distribution of docking times for the docking problem: case $k = 0.02$, $\sigma = 0.1$.

runs for a few values of σ near 1. As in Fig. 9.5, we let the vertical bars represent the range of outcomes, and the boxes connected by the dotted line indicate the average outcome for each value of σ. It seems that our overall conclusions are fairly insensitive to the exact value of σ. In every case the average docking time is around 300 seconds (5 minutes) and the variation in docking times is quite large.

Probably the most important parameter for sensitivity analysis is the control parameter k. We have assumed $k = 0.02$, which results in an average docking time of around 300 minutes. Figure 9.13 shows the results of some additional model runs in which we varied k. We made 20 additional model runs for each new value of k. As before, the vertical bars denote the range of outcomes, and the boxes represent the average outcome for each value of k. As we would expect, docking time is reasonably sensitive to the control parameter k. Of course, it is of considerable interest to determine the best value of k. We will leave this problem for the exercises.

The main robustness question for this model is the distribution of c_n, which we assumed was normal. Since our results did not vary significantly with small changes in σ, and since we are not demanding much accuracy, there is ample reason to expect the model to exhibit robustness with respect to the distribution of c_n. There are a few simple experiments to verify robustness that suggest themselves. We leave these to the exercises.

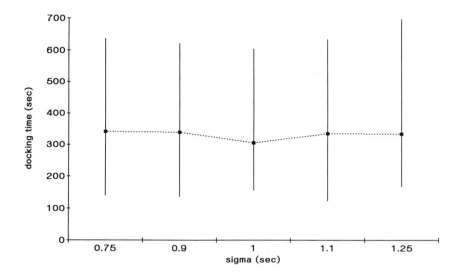

Figure 9.12 Graph of docking time versus standard deviation σ of time to make nth control adjustment for the docking problem.

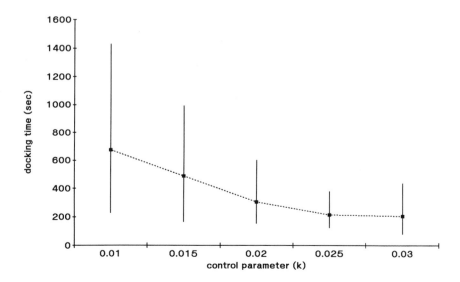

Figure 9.13 Graph of docking time versus control parameter k for the docking problem: case $\sigma = 1$.

9.3   Analytic Simulation

Monte Carlo simulation models are relatively easy to formulate, and they are intuitively appealing. Their major drawback is that a very large number of model runs is required to obtain reliable results, especially in the area of sensitivity analysis. Analytic simulations are more difficult to formulate, but are computationally more efficient.

Example 9.4   A military operations analyst plans an air strike against a well-defended target. High-altitude strategic bombers will be sent to attack this important target. It is important to ensure the success of this attack early in the battle, preferably on the first day. Each individual aircraft has a 0.5 probability of destroying the target, assuming that it can get through the air defense and then acquire (i.e., find) the target. The probability that a single aircraft will acquire the target is 0.9. The target is defended by 2 surface-to-air missile (SAM) batteries and a number of air defense guns. The flight profile of the aircraft will prevent the air defense guns from being effective (because the planes will be too high). Each SAM battery has its own tracking radar and computer guidance equipment, which is capable of tracking 2 aircraft and guiding 2 missiles simultaneously. Intelligence estimates a 0.6 probability that 1 missile will disable its target aircraft. Both SAM batteries share a target acquisition radar that is highly effective against high-altitude bombers at a range of up to 50 miles. The effective range of the tracking radar is 15 miles. The bombers will travel at 500 miles/hour at an altitude of 5 miles, and the attack requires that they loiter in the target area for 1 minute. Each SAM battery can launch 1 missile every 30 seconds, and the missiles travel at 1,000 miles/hour. How many bombers should be sent against this target to ensure its destruction?

We will use the five-step method. Step 1 is to formulate a question. We want to know how many aircraft to commit to this mission. The goal is to destroy the target, but it is immediately clear that we cannot demand a 100% guarantee of success. Let us say now that we want to be 99% sure of destroying the target. We will perform a sensitivity analysis on this number later on. Suppose that N aircraft are committed to this mission. The number of planes destroyed by air defenses before they can complete their attack on the target is a random variable. We will denote this random variable by X. In order to obtain an expression for the probability S that the mission is a success, we will proceed in two stages. First we will obtain an expression for the probability P_i of mission success given that $X = i$ planes are destroyed before they can attack the target. Then we will obtain the probability distribution of X so that we can compute

$$S = \sum_i P_i \, P_r\{X = i\}. \tag{12}$$

If N aircraft are sent and $X = i$ are destroyed before they reach the target, then there are $(N - i)$ attacking aircraft. If p is the probability that 1 attacking aircraft can destroy the target, then $(1 - p)$ is the probability that 1 attacking aircraft fails to destroy the target. The probability that all $(N - i)$ attacking aircraft fail to destroy the target is

$$(1 - p)^{N-i}.$$

Thus, the probability that at least one of the attacking aircraft succeeds in destroying the target is

$$P_i = 1 - (1 - p)^{N-i}. \tag{13}$$

The total exposure time to air defense prior to completing the attack is

$$\frac{15 \text{ miles}}{500 \text{ miles/hr}} \cdot \frac{60 \text{ min}}{\text{hr}} = 1.8 \text{ minutes}$$

on the way to the target, and an additional minute in the target area, for a total of 2.8 minutes, in which time each SAM battery fires 5 shots. Thus, the attacking aircraft will be exposed to $m = 10$ shots total. We assume the number of aircraft X destroyed will have a *binomial distribution*

$$\Pr\{X = i\} = \binom{m}{i} q^i (1 - q)^{m-i}, \tag{14}$$

where

$$\binom{m}{i} = \frac{m!}{i!(m - i)!} \tag{15}$$

is the binomial coefficient. (The distribution in Eq. (14) is the analytic model for the number of successes in m trials, where q represents the probability of success. See Exercise 12 for more details.) Now, in order to compute the probability of mission success S, we need to substitute Eqs. (13) and (14) back into Eq. (12). Then our objective is to determine the smallest N for which $S > 0.99$. This concludes step 1. We summarize our results in Figure 9.14.

Step 2 is to select our modeling approach. We will use an *analytic simulation* model. In a Monte Carlo simulation we draw random numbers to simulate events and use repeated trials to estimate probabilities and expected values. In an analytic simulation we use a combination of probability theory and computer programming to calculate probabilities and expected values. Analytic simulations are more mathematically sophisticated, which accounts for their greater efficiency. The

Variables: N = number of bombers sent
                   m = number of missiles fired
                   p = probability 1 bomber can destroy target
                   q = probability 1 missile can disable bomber
                   X = number of bombers disabled prior to attack
                   P_i = probability of mission success given $X = i$
                   S = overall probability of mission success

Assumptions: p $= (0.9)(0.5)$
                   q $= 0.6$
                   m $= 10$
                   P_i $= 1 - (1 - P)^{N-i}$
                   $\Pr\{X = i\} = \binom{m}{i}q^i(1-q)^{m-i}; i = 0, 1, 2, \ldots, m$
                   S $= \displaystyle\sum_{i=0}^{m} P_i \Pr\{X = i\}$

Objective: Find the smallest N for which $S > 0.99$

Figure 9.14 Results of step 1 of the bombing run problem.

feasibility of analytic simulation depends on both the problem complexity and the skill of the modeler. Most highly skilled analysts consider Monte Carlo simulation as a last resort, to be employed only if they are unable to formulate a suitable analytic model.

Step 3 is to formulate the model. Figure 9.15 shows an algorithm for analytic simulation of the bombing run problem. The notation Binomial (m, i, q) denotes the value of the binomial probability defined by Eqs. (14) and (15).

Step 4 is to solve the model. We used a computer implementation of the algorithm in Fig. 9.15, with inputs

$$m = 10$$
$$p = (0.9)(0.5)$$
$$q = 0.6,$$

and we varied N to obtain the results shown in Figure 9.16. A minimum of $N = 15$ planes is required to ensure a 99% chance of mission success. Step 5

Algorithm: Bombing Run Problem

Variables: N = number of bombers sent
                     m = number of missiles fired
                     p = probability 1 bomber can destroy target
                     q = probability 1 missile can disable bomber
                     S = probability of mission success

Input: N, m, p, q

Process: Begin
                     $S \leftarrow 0$
                     for $i = 0$ to m do
                         Begin
                         $P \leftarrow 1 - (1 - p)^{N-i}$
                         $B \leftarrow$ Binomial (m, i, q)
                         $S \leftarrow S + P \cdot B$
                         End
                     End

Output: S

Figure 9.15 Pseudocode for analytic simulation of the bombing run problem.

is to answer the question, which is how many bombers we must commit to this mission in order to be 99% sure of success. The answer is 15. Now we need to conduct a sensitivity analysis to get at the broader question of what would be a good number of bombers to commit to this mission.

First let us consider the desired probability of success $S = 0.99$. This is a number we literally just made up. Figure 9.17 shows the effect of varying this parameter. It appears that $N = 15$ is a reasonable decision, although any number greater than 10 and less than 20 would be fine. Sending more than 20 aircraft would be overkill.

Bad weather would decrease the detection probability, which factored into the equation $p = (0.9)(0.5)$. If the detection probability decreases to 0.5, then

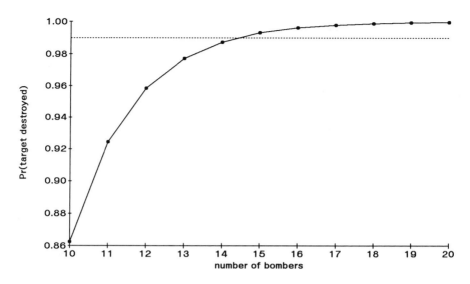

Figure 9.16 Graph of mission success probability S versus number of bombers sent N for the bombing run problem.

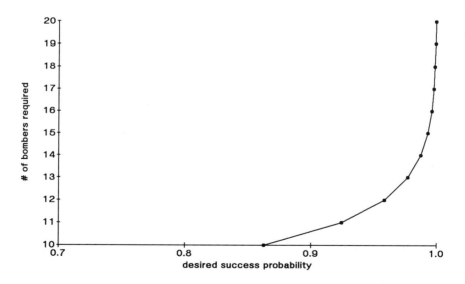

Figure 9.17 Graph of minimum number of bombers N required to obtain mission success probability S in the bombing run problem.

$p = 0.25$, and for $S > 0.99$ we will need at least $N = 23$ aircraft. If the detection probability is 0.3, then $N = 35$ aircraft are required. The overall relationship between the detection probability and the number of planes required for mission success is illustrated in Figure 9.18. It is unlikely that we will want to fly this mission in bad weather.

One of the applications for models of this kind is to analyze the potential operational impact of engineering advances. Suppose we had a bomber that flew at 1,200 miles/hour and which reduced the loitering time at the target to 15 seconds. Now the aircraft will be exposed to SAM fire for only 1 minute, so that the air defense can only shoot $m = 4$ missiles. Now $N = 11$ bombers are required to get a 99% chance of success. Figure 9.19 shows the relationship between the number of bombers sent and the probability of mission success in the baseline case ($m = 10$ missiles are fired by the air defense) and in the case of the advanced concept aircraft ($m = 4$). For the sake of comparison we also include the case $m = 0$ missiles fired. This curve represents the maximum potential benefit of the proposed technology. If bombers are exposed to no threat from air defense, it still takes at least 8 planes to ensure a 99% chance of success.

Suppose that we would produce a better targeting system that increased the probability of one aircraft destroying the target to 0.8. Now $p = (0.9)(0.8) = 0.72$ and, all else remaining the same, it would take $N = 13$ aircraft to achieve $S > 0.99$.

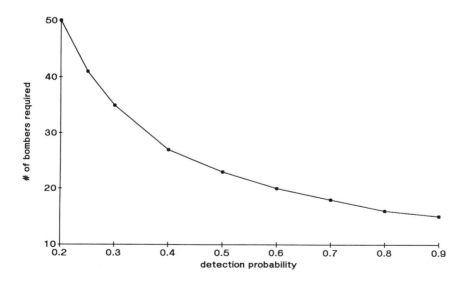

Figure 9.18 Graph of minimum number of bombers N required for mission success versus probability bombers can detect target for the bombing run problem.

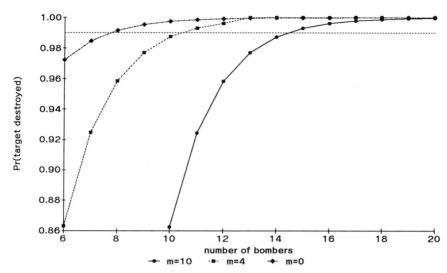

Figure 9.19 Graph of probability of mission success S versus number of bombers sent N for the bombing run problem: comparison of the three cases where the number of missiles fired is $m = 0, 4, 10$.

This is not much of an improvement. If we combine the high-speed bomber with the high-accuracy bombs, we can trim the number of planes needed to 8.

What if we have underestimated the effectiveness of enemy air defenses? If $q = 0.8$, then $N = 17$ aircraft are required for $S > 0.99$. If $q = 0.6$ but $m = 15$ (assume that there are 3 SAM batteries), then 18 planes are needed. In either case our general conclusions are not seriously altered.

On the subject of robustness, we have made a number of simplifying assumptions in our model. We have assumed that each of the attacking aircraft acquires the target independently and with the same probability. In reality the first bombs may throw up smoke and dust, obscuring the target area. This may reduce the probability that the remaining bombers can acquire the target. We have already performed a sensitivity analysis on this parameter and found that N is quite sensitive to this probability. Although our model cannot represent a dependence here, it does provide bounds. If each individual aircraft has at least a 50% chance of target acquisition, then $N = 23$ planes will suffice.

Our model also assumed that the SAM batteries never shoot twice at the same aircraft. This is certainly the optimal strategy in a target-rich environment, where the number of potential targets exceeds the maximum number of shots. But suppose that a new stealth technology is able to reduce significantly the number of aircraft detected. Now the air defense may have more shots than targets. Assume that the air defense is able to tell whether they have disabled an aircraft, so that they will not waste any shots. If d aircraft are detected, then the number of aircraft kills in m shots may be represented using a Markov chain model where

$$X_n \in \{0, 1, \ldots, d\}$$

is the number of hits after n shots. The state transition probabilities are

$$\Pr\{X_{n+1} = i + 1 | X_n = i\} = q \text{ and } P_r\{X_{n+1} = i | X_n = i\} = 1 - q$$

for $0 \leq i < d$, and of course

$$\Pr\{X_{n+1} = d | X_n = d\} = 1.$$

If $X_0 = 0$, then X_m is the number of aircraft lost. It would be possible to calculate the probability distribution of X_m for each $d = 1, \ldots, N$ and incorporate this into a modified version of our model. This is an example of the application of transient analysis for a Markov chain. It is also much easier to implement a Monte Carlo simulation for this problem. See Exercises 14 and 15 below. Our analytic simulation model is not very robust in this regard.

9.4 Exercises

1. A simple game of chance is played by flipping a coin. The house flips a coin and the player calls it in the air. If the coin lands the way the player called it, the house pays the player $1, otherwise the player pays the house $1. The player begins with $10.
    (a) What are the odds that the player will go broke before he doubles his money? Use the five-step method, and model using Monte Carlo simulation.
    (b) How long on average does the game described in part (a) last?
    (c) How much on average does the player have after 25 coin flips?
2. On a roll of 2 dice, a total of 7 occurs with probability $1/6$.
    (a) In 100 rolls of the dice, what is the probability that 5 consecutive rolls of 7 will occur? Use the five-step method, and model using Monte Carlo simulation.
    (b) What is the average number of rolls until a roll of seven occurs? Use any method.
3. Reconsider the inventory problem of Example 8.1. In the text we stated that if the time between customer arrivals is negative exponential with mean 1, the distribution of the number of arrivals in one week is Poisson with mean 1.

(a) Use Monte Carlo simulation to model one week of arrivals. Assume that the time between arrivals is negative exponential with mean 1. Simulate to determine the mean number of arrivals in one week.

(b) Modify the simulation program to keep track of the fraction of the simulated weeks in which there were 0, 1, 2, 3, or more than 3 arrivals. Compare with the Poisson probabilities given in Section 8.1.

4. (a) Repeat Exercise 3, but now assume that the time between customer arrivals is uniformly distributed between 0 and 2 weeks.

(b) Use the probabilities from your computer program output to modify the state transition probabilities for the Markov chain in Example 8.1.

(c) Solve $\pi = \pi P$ for the steady-state probabilities, assuming a uniform customer interarrival time.

(d) Determine the steady-state probability that demand exceeds supply for this modified example.

(e) Compare the results of (d) to the calculations of Section 8.1, and comment on the robustness of our original model with respect to the assumption of random arrivals.

5. Reconsider the docking problem of Example 9.3, but now assume that the time to make a control adjustment is uniformly distributed between 4 and 6 seconds.

(a) Modify the algorithm in Fig. 9.10 to reflect this change in the distribution of c_n.

(b) Implement the algorithm in part (a) on a computer.

(c) Make 20 simulation runs with $k = 0.02$, and tabulate your results. Estimate the mean time to dock.

(d) Compare the results of (c) to those obtained in Section 9.2. Would you say that the model is robust with respect to the assumption that c_n is normally distributed?

6. Reconsider the docking problem of Example 9.3.

(a) Implement the algorithm in Fig. 9.10 on a computer. Make a few model runs, and compare with the results in the text.

(b) Vary the parameter k to determine the optimal value of this control parameter. You will need to make several model runs for each value of k to determine average behavior.

(c) Explore the sensitivity of your answer in part (b) to the initial velocity of the spacecraft, which we assumed to be 50 m/sec.

(d) Explore the sensitivity of your answer in (b) to the docking threshold, which we assumed was 0.1 ft/sec.

7. Reconsider the rainy day problem of Example 9.1, but now assume that if today is rainy, there is a 75% chance that tomorrow will be rainy, and likewise

if today is sunny, there is a 75% chance that tomorrow will be sunny. The day we arrived on vacation was sunny ($X_0 = 0$).

    (a)   Determine the steady-state probability that any future day will be rainy. Model $\{X_t\}$ as a Markov chain.

    (b)   Estimate the probability of 3 consecutive days of rain using a Monte Carlo simulation.

8.   Reconsider the cell-division problem of Exercise 10, Chapter 8.

    (a)   Use Monte Carlo simulation to model the cell-division process. How long will it take on average for 1 cell to grow to 100 cells?

    (b)   What is the probability that the family line of 1 cell will die out before reaching 100 cells?

9.   (Continuation of Chapter 7, Exercise 8) Simulate a random arrival process with an arrival rate of 5 minutes using the Monte Carlo method. Determine the average time between the last arrival prior to $t = 1$ hour and the next arrival after $t = 1$ hour.

10. (Continuation of Chapter 7, Exercise 9) Simulate the supermarket checkout stand problem. Draw a random number to represent your service time, and then see how many random numbers it takes to exceed that. Repeat the simulation a large number of times and determine the average number it takes to find a random number that exceeds yours. Justify the difference between your answer and the one obtained in part (c) of Chapter 7, Exercise 9.

11. Reconsider the inventory problem of Example 8.1, and determine the average number of lost sales per week.

    (a)   Assume that on each day a customer arrives with probability 0.2. Construct a Monte Carlo simulation based on a time step of 1 day to simulate 1 week of sales activity. For a beginning inventory of 1, 2, or 3 aquariums determine the average number of lost sales by repeated simulation.

    (b)   Combine the results of part (a) with the steady-state probabilities calculated in Section 8.1 to determine the overall average number of lost sales per week.

12. This exercise explains the binomial model. Suppose m independent random trials are conducted, each of which has a probability q of success. Let $X_i = 1$ if the ith trial is successful, and $X_i = 0$ otherwise. Then $X = X_1 + \cdots + X_m$ is the number of successes.

    (a)   Show that $EX = mq$ and $VX = mq(1 - q)$. [Hint: First show that $EX_i = q$ and $VX_i = q(1 - q)$.]

    (b)   Explain why there are $\binom{m}{i}$ possible ways for $X = i$ to occur and why each one has probability $q^i(1 - q)^{m-i}$.

    (c)   Explain why Eq. (14) in the text represents the probability of i successes in m trials.

13. Reconsider the bombing run problem of Example 9.4. Modify the model to print out the expected number of aircraft lost during this mission. Take into account the possibility that additional planes are lost after the attack is concluded, as the bombers are leaving the target area.
    (a) How many planes on average are lost during the mission if $N = 15$ are sent?
    (b) Perform a sensitivity analysis with respect to N.
    (c) What happens if we can use advanced bombers that fly at 1,200 miles/hr and which only need loiter in the target area for 15 seconds?
    (d) Perform a sensitivity analysis on the probability q that 1 missile kills 1 plane. Consider $q = 0.4$, 0.5, 0.6, 0.7, and 0.8. State your general conclusions. Under what circumstances would a responsible commander order his pilots to fly this mission?

14. Reconsider the bombing run problem of Example 9.4. Suppose that superior technology allows most bombers to get through the air defense undetected.
    (a) Suppose that 4 aircraft are detected. Let Y denote the number of these aircraft who survive eight shots by the air defense. Determine the probability distribution of Y and calculate the mean number of aircraft lost prior to completion of the attack. Use a Monte Carlo simulation based on a Markov chain model, as discussed at the end of Section 9.3.
    (b) Repeat part (a), but now use an analytic simulation.
    (c) Suppose that you were required to incorporate the possibility of multiple shots at a single aircraft into the model of Example 9.3. You have two options available to you. You may write a purely analytic simulation incorporating the results of part (b), or you may use a generalized version of the Monte Carlo simulation model of part (a) to obtain the probability distribution of Y for $d = 1, \ldots, 7$ aircraft detected, and incorporate these results into the model as data. Which option would you choose? Explain.

15. (Hard problem) Carry out the model enhancements described in problem 14(c).

16. Reconsider the bombing run problem of Example 9.4.
    (a) Use Monte Carlo simulation to find the probability of mission success if $N = 15$ aircraft are sent.
    (b) Perform a sensitivity analysis on N. Determine the approximate probability of mission success for $N = 12$, 15, 18, and 21.
    (c) Compare the relative advantages of Monte Carlo and analytic simulations in terms of the difficulty of both model formulation and sensitivity analysis.

17. Reconsider the bombing run problem of Example 9.4. This problem shows how the binomial formula

$$(a + b)^n = \sum_{i=0}^{n} \binom{n}{i} a^i b^{n-i}$$

can be used to simplify the analytic simulation model presented in the text.
(a) Use the binomial formula to derive the equation

$$S = 1 - (1 - p)^{N-m}(q + (1 - p)(1 - q))^m.$$

(b) Show that the number N of planes required to ensure a success probability S is (the smallest integer greater than or equal to)

$$N = \log\left[\frac{(1 - S)(1 - p)^m}{(q + (1 - p)(1 - q))^m}\right] / \log(1 - p).$$

(c) Use this formula to verify the sensitivity analysis results reported in Fig. 9.17.

18. A radio communications channel is active 20% of the time and idle 80% of the time. The average message lasts 20 seconds. A scanning sensor monitors the channel periodically in an attempt to detect the location of emitters using this channel. An analytic simulation model of scanner performance is to be constructed. It would greatly simplify the model if it were at least approximately true that the state of the channel (busy or idle) when 1 scan is made is independent of the state found during the previous scan. Model the channel using a two-state Markov process and use an analytic simulation to determine how long it takes until the process settles down into steady-state. After this point the process essentially forgets its original state.
(a) Determine the steady-state distribution for this Markov process.
(b) Derive the set of differential equations satisfied by the state probabilities $P_t(i) = \Pr\{X_t = i\}$. [See Section 8.2.]
(c) If $X_t = 0$ (channel idle), how long does it take for the state probabilities to get within 5% of their steady-state values?
(d) Repeat part (c), assuming that $X_t = 1$ (channel busy).
(e) How far apart should successive scans be in order for the Markov property to apply (at least approximately) for this model?

19. This problem suggests a way to solve the rainy day problem of Example 9.1 using an analytic model.
(a) Let C_t denote the number of consecutive rainy days by day t. Show that $\{C_t\}$ is a Markov chain. Write down the transition diagram and the transition matrix for this Markov chain.

(b) We are interested in the probability that $\max\{C_1, \ldots, C_7\} \geq 3$. Alter the Markov chain model from part (a) by restricting the state space to just $\{0, 1, 2, 3\}$. Change the state transition probabilities so that 3 is an absorbing state, i.e., set

$$\Pr\{C_{t+1} = 3 | C_t = 3\} = 1.$$

Explain why the probability of at least 3 consecutive rainy days this week is the same as $\Pr\{C_7 = 3 | C_0 = 0\}$.

(c) Use the methods of Chapter 8 to calculate the probability of at least 3 consecutive rainy days in 1 week, assuming a 50% chance of rain each day.

(d) Perform a sensitivity analysis on the 50% assumption. Compare with the results shown in Fig. 9.5.

(e) Compare this analytic model to the Monte Carlo model used in Section 9.1. Which do you prefer, and why? If you had just now come across this problem, which modeling approach would you have selected?

Further Reading

Bratley, P. et al. (1983). *A Guide to Simulation*, Springer-Verlag, New York.

Hoffman, D. *Monte Carlo: The Use of Random Digits to Simulate Experiments*, UMAP module 269.

Meerschaert, M. and Cherry, W. P. (1988). Modeling the behavior of a scanning radio communications sensor, *Naval Research Logistics Quarterly*, Vol. 35, 307–315.

Molloy, M. (1989). *Fundamentals of Performance Modeling*, Macmillan, New York.

Press, W. et. al. (1987). *Numerical Recipies*, Cambridge University Press, New York.

Rubenstein, R. (1981). *Simulation and the Monte Carlo Method*, Wiley, New York.

Shephard, R., Hartley, D., Haysman, P., Thorpe, L. and M. Bathe. (1988). *Applied Operations research: Examples from Defense Assessment*, Plenum Press, London.

Afterword

Mathematics is the language of problem solving. It is at the heart of all science and technology. The beauty of an education in mathematics is that it gives you the freedom to pursue just about any technical career you can imagine. In the next few pages we will provide a brief discussion of some of the more common career opportunities for mathematics majors. We will also provide some suggestions about how to achieve success in a job that uses mathematics to solve real-world problems. The number-one question in the mind of most students is whether to go directly to work, or whether to pursue an advanced degree. We will begin by describing the most plentiful job opportunities for bachelor's and master's degree mathematics graduates.

Computer programming is currently the number-one job opportunity for mathematics majors. There is a strong and steadily growing demand for competent programmers which is expected to continue for the foreseeable future. Mathematics majors are better-than-average programmers, and employers know this. All you need to do to qualify for most entry-level jobs is to complete three or four courses in computers, including a course in data structures, and have knowledge of a commonly useful programming language like C, FORTRAN, or COBOL. A degree in computer science is not usually required, simply because there are not enough computer science graduates to fill the available jobs. Computer programming is an easily marketable skill that can open many doors for you. After you get a job and prove yourself as a programmer, you will find that many other opportunities present themselves.

Another good job opportunity for mathematics graduates is actuarial work. Actuarial firms will often hire a good mathematics major just on the basis of his of her academic record, but if you are really interested, it is a good idea to take the first actuarial exam (covering calculus and linear algebra) before you graduate. To become a full-fledged actuary requires passing a series of 10 exams. There are graduate courses in actuarial science which will help you to prepare for the exams, or you can study on your own. If you have the ability and self-discipline, you can rise to the top of a very interesting and lucrative profession in 10 years or less. Almost every Fortune 500 company has at least one vice president who is an actuary, not to mention the opportunities at insurance companies and independent actuarial firms. Actuaries do the mathematical modeling for these firms. This is an especially good track for people with a strong interest in business, but you do not need to have any course work in business, economics, or accounting to qualify.

Many students choose to combine their degree in mathematics with a second degree in engineering, computer science, or accounting. Of course, a good student in one of these areas will have no problem landing a respectable job. One question he or she may have is to what extent that job will use the full measure of his or her knowledge in mathematics. To be honest, there is good news and bad news here. The bad news is that during the first year or two of employment, there may be little opportunity to use much college-level mathematics. Everybody has to start somewhere, and most people have to start at the bottom. The important thing to remember is to perform with energy, enthusiasm, and accuracy. Look at this part of your working career as a test to see whether you are ready for a more sophisticated job. The good news is that once you get past the opening stanza, there is a wealth of genuine opportunities to put your math to work on significant and interesting problems. Of course, you will have to prove that you can handle the challenge.

Next we will discuss some of the opportunities for mathematics graduates who intend to pursue an advanced degree. Mathematics majors are welcome in virtually every graduate program, especially in science and engineering. When you look at the kind of work being done at the advanced levels in most of these areas, you will find that it involves a lot of mathematics. There is not enough room here to describe the incredible variety of opportunities available. We will concentrate on those fields primarily concerned with the mathematical aspects of solving real-world problems.

Advanced graduate education in mathematics is an obvious choice for someone with a degree in mathematics. If you are interested in solving real-world problems, you should look for a graduate school that offers a program in applied mathematics, statistics, or operations research. Computer science is another attractive choice, for reasons stated earlier. Bright mathematics majors with a good background in computing will find ample opportunity to use their mathematical skills during graduate school in computer science. After graduation these folks will possess a powerful combination of mathematical and computer skills which can be brought to bear on a wide variety of fascinating real-world problems. Statistics is another good option. Most statisticians started out by earning a degree in mathematics.

There are plenty of job openings, and the work is much more varied than most people believe. Some of the most interesting work in mathematical modeling is done by statisticians.

Operations research is a very broad field of study that encompasses most of what we usually think of as mathematical modeling. It includes the study of problems in optimization, queuing theory, and inventory theory. It is possible to enter the field with a degree in mathematics, but it is better to start with an advanced degree in the field of operations research. The biggest problem here is to find the right program. Mathematics, statistics, computer science, engineering, and even MBA programs often offer a major or a concentration in this area. Choose any one you like; it does not make too much difference which department grants the degree. Different schools have different philosophies about where such a program belongs. Another confusing problem is the name of the program. Operations research, operations management, management science, and systems science are all different names for essentially the same thing. Once again, it does not really matter which one of these appears on your degree.

If you are interested in research and teaching, consider a doctoral program. A doctorate in mathematics is one of the more marketable degrees in academics, and a doctorate in some branch of applied mathematics (e.g., numerical analysis or partial differential equations) can lead to a very good job in an industry research laboratory. If you are interested in an academic job, you should also consider earning your doctorate in statistics, computer science, or operations research. You will find that the work is mostly math, the job market is better, and the salaries are higher. In fact, a doctoral program in almost any field of science or engineering can provide a wealth of opportunities to use mathematics to solve real-world problems. You should not be afraid to pursue anything that captures your imagination. Finally, do not overlook the possibility of a combined doctorate. Programs in mathematical physics, mathematical biology, mathematical psychology, and mathematical economics offer unique challenges. You may also want to begin thinking about the choice between an academic job and an industry job. While academic jobs offer much in the way of lifestyle benefits, industry jobs typically pay about twice as much. You will want to be in a position to choose.

Further Reading

The Actuarial Profession, Society of Actuaries, 475 North Martingale Road, Schaumburg IL 60173–2226.

Careers in Applied Mathematics, Society for Industrial and Applied Mathematics, 3600 University City Science Center, Philadelphia PA 19104–2688.

Careers in the Mathematical Sciences, Mathematical Association of America, 1529 18th St. NW, Washington DC 20036.

Careers in Operations Research, Operations Research Society of America, Mount Royal and Guilford Aves., Baltimore MD 21202–9990.

Careers in Statistics, American Statistical Association, 1429 Duke St., Alexandria VA 22314–3402.

Computer and Mathematics Related Occupations, Bulletin 2300, US Department of Labor, Bureau of Labor, Statistics Publication Sales Center, PO Box 2145, Chicago IL 60690.

Index